VIRTUAL REALITY

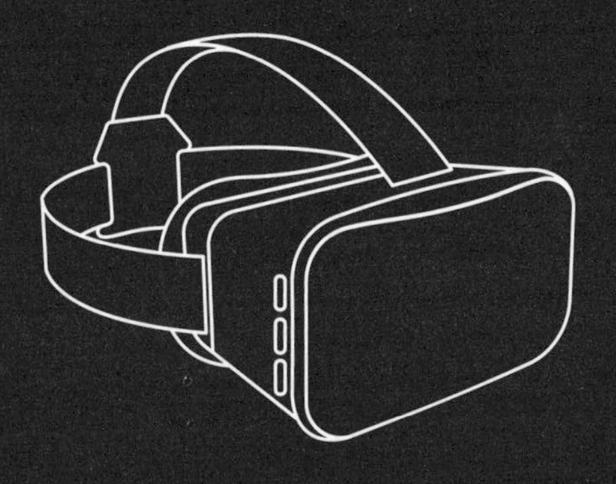

The MIT Press Essential Knowledge Series

A complete list of the titles in this series appears at the back of this book.

VIRTUAL REALITY

SAMUEL GREENGARD

The MIT Press | Cambridge, Massachusetts | London, England

This book was set in Chaparral Pro by Toppan Best-set Premedia Limited. Printed and bound in the United States of America.

Library of Congress Cataloging-in-Publication Data

Names: Greengard, Samuel, author.
Title: Virtual reality / Samuel Greengard.
Description: Cambridge, MA : The MIT Press, 2019. | Series: The MIT Press essential knowledge series | Includes bibliographical references and index.
Identifiers: LCCN 2018058335 | ISBN 9780262537520 (pbk. : alk. paper)
Subjects: LCSH: Virtual reality—Popular works.
Classification: LCC QA76.9.C65 G745 2019 | DDC 006.8—dc23 LC record available at https://lccn.loc.gov/2018058335

10 9 8 7 6 5 4 3 2 1

CONTENTS

SERIES FOREWORD

The MIT Press Essential Knowledge series offers accessible, concise, beautifully produced pocket-size books on topics of current interest. Written by leading thinkers, the books in this series deliver expert overviews of subjects that range from the cultural and the historical to the scientific and the technical.

In today's era of instant information gratification, we have ready access to opinions, rationalizations, and superficial descriptions. Much harder to come by is the foundational knowledge that informs a principled understanding of the world. Essential Knowledge books fill that need. Synthesizing specialized subject matter for nonspecialists and engaging critical topics through fundamentals, each of these compact volumes offers readers a point of access to complex ideas.

Bruce Tidor
Professor of Biological Engineering and Computer Science
Massachusetts Institute of Technology

ACKNOWLEDGMENTS

The reality of writing a book is formidable. *Virtual Reality* was no exception. Quite a few people contributed to this project in one form or another.

First, I'd like to thank MIT Press executive editor Marie Lufkin Lee, who approached me about writing another book for the Essential Knowledge series. We kicked around several ideas before deciding that extended reality warranted attention. I never regretted the decision, even when I was up to my eyeballs in research about complex technical concepts and sinking under the weight of detailed interviews with experts.

I'd also like to thank several people who aided in developing content for this book. Accenture managing director Peter Soh and media relations specialist Hannah Unkefer extended an invitation to an Accenture showcase in February 2018. It highlighted leading-edge virtual-, augmented-, and mixed-reality systems and apps. They also provided access to several Accenture subject matter experts at the firm, including Marc Carrel-Billiard, Emmanuel Viale, Mary Hamilton, and Jason Welsh. These subject matter experts provided key ideas and concepts that helped frame the direction of the book.

I'm also deeply grateful to Albert "Skip" Rizzo at the University of Southern California, who offered valuable insights into psychology, physiology, and the effects virtual reality has on humans, and Jeremy Bailenson at Stanford University, a thought leader in the field of virtual reality who provided key guidance about the topic. Thanks also to Jake Rubin, founder and CEO of HaptX, a company on the leading edge of VR systems development.

Of course, a book can't take shape without editors. I'd like to tip my cap to Stephanie Cohen at MIT Press, who shepherded the project through to completion. She was extremely easy to work with and always available to answer questions or deal with issues. I'd also like to acknowledge the diligence of Michael Sims at MIT Press, who caught numerous minor errors and typos while proofreading the manuscript. And thanks, too, to the four anonymous reviewers who provided valuable feedback about concepts and spotted a few omissions and errors along the way. Their comments and suggestions ultimately led to a much better book.

I'd also like to thank my partner, Patricia Hampel Valles, for enduring ongoing discussions about virtual and augmented reality—and for providing thoughts about key ideas and concepts that appear in the book. Finally, a shout out to my sons, Evan and Alec, who are emerging as smart young men. Both are now attending university and learning that knowledge is far more than the sum of YouTube videos and Wikipedia entries.

INTRODUCTION

Standing at the edge of an abyss that drops thousands of feet probably wouldn't rank as a bucket-list item for most people. It's more like a terrifying moment to avoid at any and all costs. But this doesn't change the fact that I must walk across a narrow ramp and board a spaceship—with no time to spare. At this moment, my mental state could be described as somewhere between absolute bliss and sheer terror. My heart is racing while a queasy sensation has invaded the pit of my stomach. My brain, my eyes, and my senses tell me that with one wrong step I will plummet into oblivion. So, I take tiny and careful steps—one ... at ... a ... precious ... time—until I have managed to traverse the narrow ramp. As I enter the space vehicle I hear myself emit a sigh of relief.

Of course, I'm not standing in a real spaceship. And I didn't cross an actual boarding ramp. Truth be told, I'm not even remotely close to outer space. I'm wandering through an immersive virtual-reality environment called The VOID, which is located in Las Vegas. Here, a visitor fastens a vest and gloves, dons a virtual-reality head-mounted display, and ventures into a holodeck—an environment that blends physical and virtual reality. I've just entered a Star Wars battle called Secrets of the Empire.

Once inside this space, the bare walls, floors, and ceilings of the holodeck morph into the spaceship. I'm actually a participant in a Star Wars film. It's a 3D world where I can feel heat and vibrations, smell damp air and smoke, hear the sounds of battle, and, using my proton blaster, a laser gun, take out enemy storm troopers.

The experience, which lasts about 20 minutes, is at times exhilarating and at other times completely terrifying. Throughout the adventure, however, it's entirely convincing. The VOID is unlike any form of entertainment I've ever experienced. It is to virtual reality what recordings are to music or smartphones are to communication. The space seamlessly weaves together reality and fantasy to advance the boundaries of entertainment to a new frontier. Forget the flat screens and 2D experience that TVs, computers, and movies deliver. This immersive world seems completely and utterly real. The VOID website describes the environment as "hyperrealism."

The VOID represents the leading edge of a massive wave of virtual reality, augmented reality, and mixed reality. In family rooms, offices, and research labs around the world, people are using special goggles or head-mounted displays—along with haptic gloves and other apparatus—to venture inside a computer-generated world that convincingly mimics our physical world. In addition, many are turning to augmented-reality eyeglasses and smartphone apps to view data, graphics, and images in new and intriguing ways. These capabilities are available on demand at the push of a button or via a voice command.

To be sure, after decades of hype and sometimes breathless predictions, extended reality is taking shape. Zion Market Research estimates that the total market for virtual reality will swell from $2.2 billion in 2016 to $26.89 billion in 2022.[1] According to the market research firm ARtillry Intelligence, the augmented-reality market will reach $18.8 billion by 2022.[2] Yet these technologies are also reshaping business. The number of companies entering the augmented-reality market is growing at an annual rate of about 50 percent.[3] Total global extended reality revenues will reach $61 billion in 2022, ARtillry reports.[4]

Yet this isn't simply a tale of numbers and adoption. Jeremy Bailenson, a professor of communications and founding director of Stanford's Virtual Human Interaction Lab has stated that these technologies have significant psychological and social ramifications.[5] Once a person steps inside a virtual world, things change. "VR [virtual reality] takes all the gadgets away, it takes all the multitasking away and you actually feel like you're with someone. We call this social presence—you see their emotions, you see their gestures and it feels just like you're in the room with them. It takes what is typically seen as something that's unemotional and distant and makes it feel like somebody's right there with you."

It's impossible to argue with Bailenson's conclusion. When I strap on an Oculus Go headset, reality is transformed. I can tour the canals of Venice, Italy, and turn my head to view things—boats, people, and buildings—as if I actually am taking a gondola ride. I can step aboard the International Space Station, learn about the different modules and equipment, and view the earth below. I can drop into fantasy worlds and games and even feel as though I'm on the court during a professional basketball game. It's possible to watch LeBron James dunk or Stephen Curry throw down a three-point shot as if I were actually there. Even the most advanced television—delivering an ultra-high definition experience—cannot duplicate the feeling.

AR glasses also transform the way I see the world. I receive data and information in visual and auditory formats that are more useful, more contextual, and more refined than any laptop or smartphone can deliver. Suddenly, it's possible to view a video that shows how to install a light switch—*while I'm actually installing the light switch.* With both hands free, the task is much easier. There's no switching back and forth between a video playing on my smartphone and tinkering with the physical light switch—while juggling tools. This explains why architects, engineers, scientists, insurance agents, and countless others have begun to use artificial reality glasses to tackle a vast array of tasks.

There's also mixed reality, which refers to a world somewhere between virtual and augmented reality. In this space, a head-mounted display projects a virtual world but includes actual physical objects in the virtual experience. Or it changes those physical objects into different things, which is what takes place inside The Void. A plain door in the holodeck becomes an elaborate control panel for entering a space ship. A rail or seat that actually exists in the room appears in the virtual world. Essentially, real world objects and virtual objects appear together in a "mixed" reality.

Haptics technologies—which intersect with virtual and augmented reality—ratchet up capabilities further. At a technology showcase presented by the global professional services firm Accenture in San Francisco in early 2018, I donned state-of-the-art haptics gloves along with an HTV Vive headset. The HaptX gloves were large and a bit clunky—they look like a robotic hand—but once I ventured into the virtual space the gloves become invisible. Suddenly, I could feel objects, including animals, shafts of wheat, and clouds. When I poked a cloud with a finger I could feel raindrops. When I picked up a rock I could feel the shape and texture. Venturing inside this odd, animated world was unlike anything I have experienced before.

Although various forms of virtual and augmented reality have appeared in books and films for a few decades, the vision of how they will play out is also evolving. In

2018, Steven Spielberg released the film *Ready Player One*. Set in the year 2045, human life on earth is threatened by severe climate change, which has unleashed a global energy crises, famine, disease, poverty, and war. The story, based on the 2011 science fiction novel of the same name, tracks protagonist Wade Watts as he escapes the unbearable reality of the physical world by traveling into a virtual world—along with millions of others who are in search of an enormous prize.

Let's not assume, however, that extended reality is a virtual magic carpet ride to a utopian place. Myriad concerns swirl around morality, ethics, legality, and social considerations. Even more important is the fundamental question whether these technologies will actually lead to a net improvement in our lives—or whether they merely represent a faster or more enjoyable way to do things. I find that if I wear a headset for more than about 45 minutes, I begin to feel isolated and anxious. And I wonder, at a time when we already engage in separate activities together—fiddling with our phones at a gathering with family and friends and everyone barely talking, for example—how things will play out when we wear a head-mounted display. At that point, watching a movie or participating in a game loses the direct physical connection between humans. It's simply a group of people doing something together while separated. Each person is isolated in his or her own virtual-reality world.

As society introduces virtual technologies that bridge the physical and digital realms, there are important questions to ask. *Will extended reality create a better world? Will it benefit society as a whole? Or will XR merely fuel financial gains for a few winners—particularly the companies and individuals that build or use these systems effectively?*

This book examines how the three types of extended reality—augmented reality, virtual reality, and mixed reality—have an impact on the world around us and how they are positioned to change the future. We will take a look at the history of these technologies, the latest research in the field, and how they are shaping and reshaping various professions and industries—along with the way we consume news, entertainment, and more. We will also examine the impact extended reality has on psychology, morality, law, and social constructs. Finally, we will speculate about how a virtual future will take shape. One thing is clear: virtual reality, augmented reality, and mixed reality have evolved from niche technologies, often relegated to gaming, to mainstream platforms that are radically changing the way we think about computing—and the way we use devices.

These technologies now appear at airports, in schools, in vehicles, on television sports broadcasts, in medical offices, and at home. Retailers now offer smartphone apps that allow consumers to preview a color of paint on the walls of their house or what a sofa will look like in their living room. There's also immersive 3D porn and a pastor who is looking to create a virtual-reality church congregation. Make no mistake, the possibilities are limited only by our imaginations.

The journey into extended reality is just getting started.

1

WHY AUGMENTED AND VIRTUAL REALITIES MATTER

The Physical and Digital Worlds Collide

The idea that perceptions can be altered is nothing new. Over the centuries, artists, inventors, and magicians have produced sometimes intriguing illusions that trick the human eye—and the mind. The first extended reality most likely appeared in the form of cave drawings and petroglyphs, a fact that Howard Rheingold pointed out in his seminal 1991 book, *Virtual Reality: The Revolutionary Technology of Computer-Generated Artificial Worlds—And How It Promises to Transform Society*. Essentially, someone sketched an image of a bison, saber-toothed cat, or person on a rock wall.

Centuries later, graphic artists began to experiment with optical illusions. For example, in 1870, German physiologist Ludimar Hermann drew a white grid on a black background. As an individual's eyes scan across the illustration, the intersection points—essentially dots—change back and forth from white to gray (see figure 1.1). In the 1920s, M. C. Escher, a graphic artist from the Netherlands, began drawing pictures that delivered physical impossibilities, such as water traveling uphill. His art is still popular today.

What Hermann, Escher, and many others understood is that the brain can be tricked into believing things—or seeing things—that aren't there or don't necessarily make logical sense. The right stimulus and sensory input can seem just as convincing as reality—or alter the way we see reality. As early as the 1830s, inventors began tinkering with stereoscopes that used optics and mirrors along with a pair of lenses to produce a 3D view of objects. The company View-Master commercialized the concept when it introduced a hand-held stereoscopic viewer in 1939 (ironically, the company is now venturing into the VR space with special goggles and software). The system presented 3D images of things and places around the world—from the Grand Canyon to Paris, France.[1] It relied on cardboard disks with pairs of embedded photographic color images to produce a more realistic feeling of "being there."

Yet, it is only since the introduction of digital computing that the concept of extended reality (XR) has

Figure 1.1 The Hermann grid was created by physiologist Ludimar Hermann. When a viewer looks at any given dot, it turns white. However, when a person looks away from a dot it shifts between white and gray. Source: Wikipedia Commons.

Virtual reality creates an illusion that a person is in a different place. Perhaps it's a meeting with business colleagues scattered all over the globe. Or perhaps it's parachuting out of an airplane, riding a roller coaster, or navigating through the canals of Venice, Italy.

emerged in the way we think about it today. Computing systems—comprising a variety of digital components and software—deliver convincing images, sound, feel, and other sensory elements that alter the way we experience existing physical things (augmented reality, or AR) or create entirely imaginary but realistic seeming worlds (virtual reality, or VR). Either way, extended reality technologies allow us to step beyond, dare we say, the limitations of the physical world and explore places only imagination could go in the past.

Today, AR and VR, along with mixed reality (MR), which simultaneously blends elements of the physical world with virtual or augmented features, are appearing in all sorts of places and situations. They're in movies, on gaming consoles, on smartphones, in automobiles, and on glasses and head-mounted displays (HMDs). They are transforming the world around us one click, tap, or glance at a time. A convergence of digital technologies, along with remarkable advances in computer power and artificial intelligence (AI), is delivering AR and VR into new and often uncharted territory.

Smartphone apps use a camera and AR to recognize physical things and display names, labels and other relevant information on the screen. They tackle real-time language translation. They show what makeup or clothing looks like on a person. It's even possible to see wine labels come alive with an animated display[2] or show what a room will look like with a particular piece of furniture or a different color scheme.[3]

At the same time, VR is appearing in games, research labs, and industrial settings that use headsets, audio inputs, haptic gloves, and other sensory tools to generate ultrarealistic sensations. Over the coming decade and beyond, these systems will change countless tasks, processes, and industries. They will also dramatically alter interactions between people through the use of *telepresence* and *telexistence*. The former term refers to systems that allow people to feel "present" when they are physical separated. The latter word revolves around the concept that a person can be in a place separate from his or her physical presence.[4]

Perhaps it's a meeting with business colleagues scattered all over the globe. Or perhaps it's parachuting from an airplane, riding a roller coaster, or rafting the rapids of a raging river. But VR is more than the illusion of being in a different place or time. These worlds can incorporate virtual artifacts (VA) that mimic objects from the physical world or things identifiable only to computers, including digital tokens that trigger certain events. For example, a person might purchase something or get paid using a digital currency such as Bitcoin when a specific automated action takes place, or a virtual object is used in a certain way.

At this point, let's define the key terms you will encounter throughout this book. *Merriam Webster's* dictionary describes augmented reality as "an enhanced version of reality created by the use of technology to overlay digital information on an image of something being viewed through a device (such as a smartphone camera).[5] It's important to note that this process takes place in real time. *Merriam Webster's* defines virtual reality as "an artificial environment which is experienced through sensory stimuli (such as sights and sounds) provided by a computer and in which one's actions partially determine what happens in the environment."[6] Mixed reality, by comparison, juxtaposes real world objects and virtual objects within a virtual space, or on AR glasses. This might mean projecting a virtual dog in a real living room, complete with actual furniture, that's viewable through a smartphone or glasses. Or it might mean projecting a real dog into a virtual world, filled with both real and imaginary objects. Augmented and mixed realities are similar—and sometimes the same. In a basic sense, it's best to think of augmented reality as supplemental and mixed reality as a mashup of both real and virtual things.

XR technologies come in many shapes and forms. Virtual reality can incorporate nonimmersive spaces such as surrounding LCD panels where only some of a user's senses are stimulated; semi-immersive spaces, like a flight simulator, that combine physical and virtual elements in a room; and fully immersive simulations that block out the physical world. Not surprisingly, the latter produces a far more realistic and engaging experience—but it also requires sophisticated hardware and software to produce a high-resolution sensory experience. Immersive VR typically includes a head-mounted display and other input and output devices, such as haptic gloves.

One way to think about extended reality is to consider VR an immersive experience and AR a complementary experience. When elements of VR and AR overlap with the physical world, the result is MR. Apps such as SnapChat and Facebook—and gaming platforms like the PlayStation and Xbox—already use virtual, augmented, and mixed reality. In the coming years, XR will expand digital interaction further. These technologies will transform static 2D environments displayed on screens into realistic and sometimes lifelike 3D representations. Moreover, XR technologies will intersect with other digital tools to produce entirely new features, capabilities, and environments.

Extended reality will profoundly change the way we connect to others in the world around us. Films such as *Tron*, *The Lawnmower Man*, *Minority Report*, *The Matrix*, and *Iron Man* showcased what's possible—or at least imaginable. In the future, various forms of extended reality will deliver us to places that no 2D screen, however sophisticated, can display. As Marc Carrel-Billiard, global senior

“The human brain ... is wired to grasp events in 3D. Extended reality further bridges the gap between humans and computers.”—Marc Carrel-Billiard, Global Senior Managing Director at Accenture Labs managing director of Accenture Labs puts it: “The human brain … is wired to grasp events in 3D. Extended reality further bridges the gap between humans and computers.”

The Birth of Extended Reality

The path to modern AR and VR has taken plenty of twists and turns. In the 1780s, Irish-born painter Robert Barker began experimenting with the idea of creating a more immersive experience. The Leicester Square panorama, the first grand display of the concept, opened in London in 1793. It featured a 10,000-square-foot panorama and a smaller 2,700-square-foot panorama within a large viewing structure.[7] It was hailed as a completely new and revolutionary idea by reviewers. By the early part of the nineteenth century, artists began creating elaborate 360-degree panoramas that produced a more realistic “virtual” sensory experience for a variety of scenes, including battles, landscapes, and famous landmarks.

In 1822, two French artists, Louis Daguerre and Charles Marie Bouton, introduced a new wrinkle: the diorama.[8] Their original creations used material painted on both sides of a screen or a backdrop. When illumination changed from front to back or side, the scene would appear differently. For example, a daytime scene would become nighttime or a train would appear on a track and look as

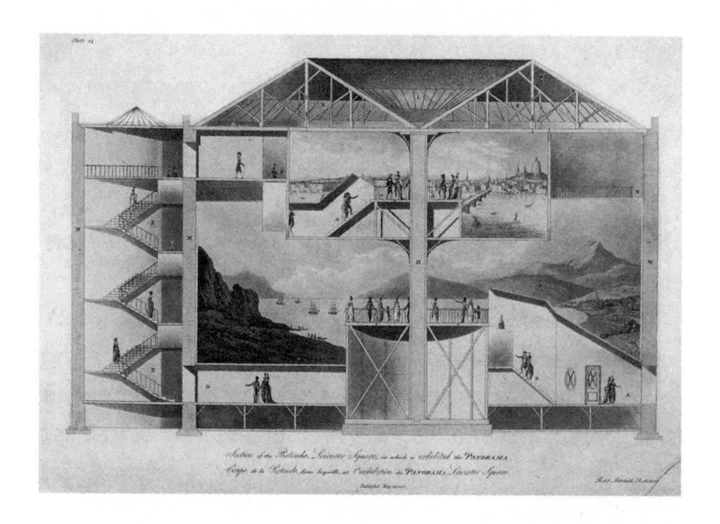

Figure 1.2 Cross-section of the Rotunda in Leicester Square, one of the first and most elaborate panoramas. By Robert Mitchell, via Wikimedia Commons.

though it has crashed. Dioramas are still used today in museums to depict natural scenes. For example, a backdrop might depict the Serengeti in Tanzania with a lion and zebra along with realistic looking plants in the foreground. The environment blends together to create a 3D illusion of actually being there.

In 1932, Aldous Huxley, in the novel *Brave New World,* introduced the idea of movies that would provide sensation, or "feelies," to transform a scene. By 1935, science fiction author Stanley G. Weinbaum framed the concept in a more tangible and modern way. In a story titled "Pygmalion's Spectacles,"[9] he served up the notion of a person wearing goggles that could depict a fictional world. The lead character in the story, Dan Burke, encountered an inventor named Albert Ludwig, who had engineered "magic spectacles" capable of producing a realistic virtual experience, including images, smell, taste, and touch.

"Pygmalion's Spectacles" may have been the world's first digital love story. It opens with Burke uttering the line: "But what is reality?" Ludwig replies: "All is dream, all is illusion; I am your vision as you are mine." Ludwig then offers Burke a movie experience that transcends sight and sound. "Suppose now I add taste, smell, even touch, if your interest is taken by the story. Suppose I make it so that you are in the story, you speak to the shadows, and the shadows reply, and instead of being on a screen, the story is all about you, and you are in it. Would that be to make real a dream?"

These imaginary spectacles produced a complete sensory experience, including sight, sound, smell, and taste. With a person fully engaged, the "mind supplies" the sensation of touch, Ludwig explained. Weinbaum described the spectacles as "a device vaguely reminiscent of a gas mask. There were goggles and a rubber mouthpiece." Later in the story, Burke winds up in an imaginary world filled with a forest and a beautiful woman named Galatea. However, the eerie and elusive world disappears like a dream and leaves him longing for the imaginary woman. "He saw finally the implication of the name Galatea ... given life by Venus in the ancient Grecian myth. But his Galatea, warm and lovely and vital, must remain forever without the gift of life, since he was neither Pygmalion nor God."

While authors like Weinbaum conjured up wild ideas about how a virtual world might look and feel, inventors were tinkering with electronic components that would serve as the humble origins of today's AR and VR. In 1929, Edward Link introduced the Link Trainer, a primitive version of a flight simulator.[10] In 1945, Thelma McCollum patented the first stereoscopic television.[11] Then, on August 28, 1962, Morton Heilig, a philosopher, filmmaker, and inventor, introduced a device called the Sensorama Simulator, which he described as an "experience theater." It transformed the concept of a virtual world into something that could be produced by an actual machine (see figure 1.3).[12]

Heilig had actually invented the Sensorama Motion Picture Projector and the 3D Motion Picture Camera in 1957.[13] But in 1960 he added a component that would transform disparate technologies into a working system. The Telesphere Mask, a head-mounted display, produced stereoscopic images, wide vision, and stereophonic sound (see figure 1.4). Four years later Heilig submitted a series of drawings and notes to the US Patent Office. Essentially, a person would sit on a chair with his or her head extending

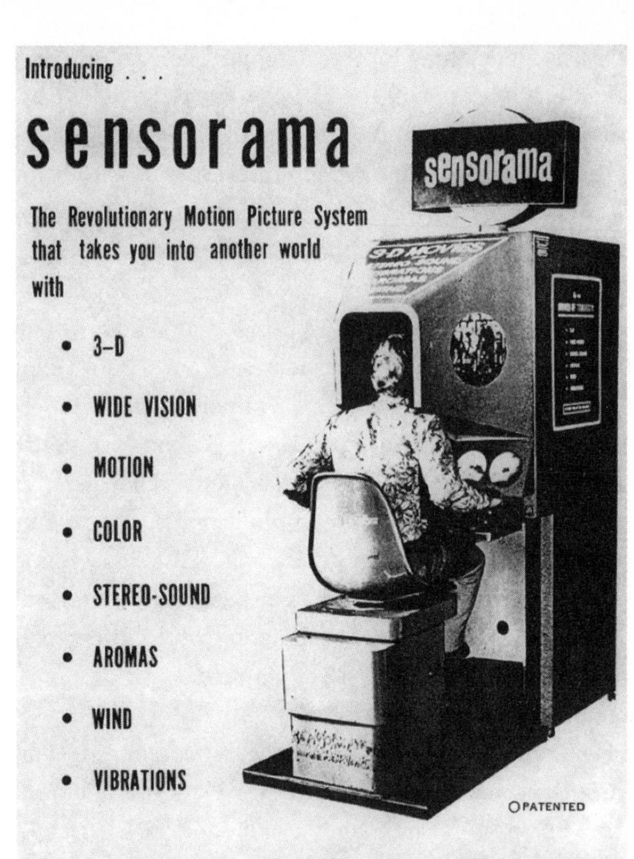

Figure 1.3 The Sensorama represented the first attempt to create a multisensory virtual-reality environment. Source: Wikipedia Commons.

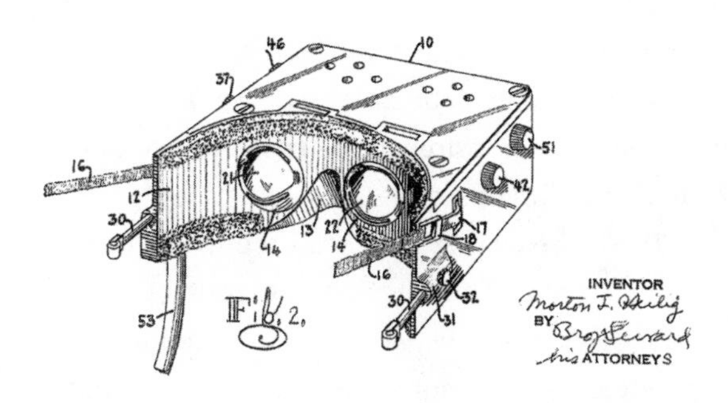

Figure 1.4 An early head-mounted display invented by Morton Heilig. The patent for the Telesphere mask was filed in 1957. Source: Wikipedia Commons.

into a surrounding apparatus that produced the virtual environment. In addition to presenting 3D visual images projected onto the "hood," the Sensorama would deliver moving air currents, various smells, binaural sound, and different types of vibrations, jolts, and other movements. Heilig created five short films for the Sensorama. They appeared as moving 3D images.

The Sensorama was a bold but unproven idea. Heilig succeeded in stepping beyond previous efforts to use multiple projectors and systems that filled only about 30 to 40 percent of a person's visual field. "The present invention [aims to] stimulate the senses of an individual to simulate an actual experience realistically," Heilig wrote in the original patent filing. "There are increasingly demands for ways and means to teach and train individuals without actually subjecting individuals to possible hazards of particular situations." The overall goal, he noted, was to "provide an apparatus to stimulate a desired experience by developing sensations in a plurality of senses."

Then, in 1961, the world's first head-mounted display (HDM) appeared. Philco Corporation, a manufacturer of electronics and televisions, began exploring the idea of a helmet that would use a remote-controlled closed-circuit video system to display lifelike images. The system, called Philco Headsight,[14] relied on head movements to monitor how a person moved and reacted. The device adjusted the display accordingly. Soon, other companies, including Bell Helicopter Company, began exploring the use of a head-mounted display along with an infrared camera to produce night vision. This goal of this AR tool was to help military pilots land aircraft in challenging conditions.

During the same period, the field of computer graphics was beginning to take shape. In 1962, Ivan Edward Sutherland, while at MIT, developed a software program called the Sketchpad, also referred to as the Robot Draftsman.[15] It introduced the world's first graphical user interface (GUI), which ran on a CRT display and used a light pen and control board. The technology later appeared in personal computers—and spawned computer aided design (CAD). Sketchpad's object-centric technology and 3D computer modeling allowed designers and artists to render convincing representations of real things. Although it would be years before researchers would connect the dots between Sketchpad and XR, Sutherland's invention was nothing less than groundbreaking.

His contributions didn't stop there. In 1965, while an associate professor at Harvard University, Sutherland penned a seminal essay about augmented and virtual realities. It essentially served as the foundation for everything that would follow. He wrote:

> We live in a physical world whose properties we have come to know well through long familiarity. We sense an involvement with this physical world which gives us the ability to predict its properties well. For example, we can predict where objects will fall, how well-known shapes look from other angles, and how much force is required to push objects against friction.
>
> The ultimate display would, of course, be a room within which the computer can control the existence of matter. A chair displayed in such a room would be good enough to sit in. Handcuffs displayed in such a room would be confining, and a bullet displayed in such a room would be fatal. With appropriate programming such a display could literally be the Wonderland into which Alice walked. [16]

Then, in 1968, extended reality took a giant leap forward. Sutherland, along with a student researcher, Bob Sproull, invented a device called the Sword of Damocles (figure 1.5).[17] The head-mounted display connected to a device suspended from the ceiling to stream computer-generated graphics to special stereoscopic glasses. The system could track a user's head and eye movements and apply specialized software to optimally adjust the images. Although the system did not provide a high level of integration between different components, it served as a proof point for future HMDs and goggles.

Morton Heilig was back at it again as the decade wound down. In 1969, he patented the Experience Theater,[18] a more sophisticated version of the Sensorama Simulator. It consisted of a motion picture theater with a large semispherical screen displaying 3D motion pictures along with speakers embedded in each chair. The system included peripheral imagery, directional sound, aromas, wind, temperature variations, and seats that tilted. "The invention relates to an improved form of motion picture or television entertainment in which the spectator is enabled, though substantially all of his senses, to experience realistically the full effect or illusion of being a part of or physically responding to the environment depicted in the motion picture," he wrote in United State Patent Office filing 3,469,837, submitted on September 30, 1969.[19]

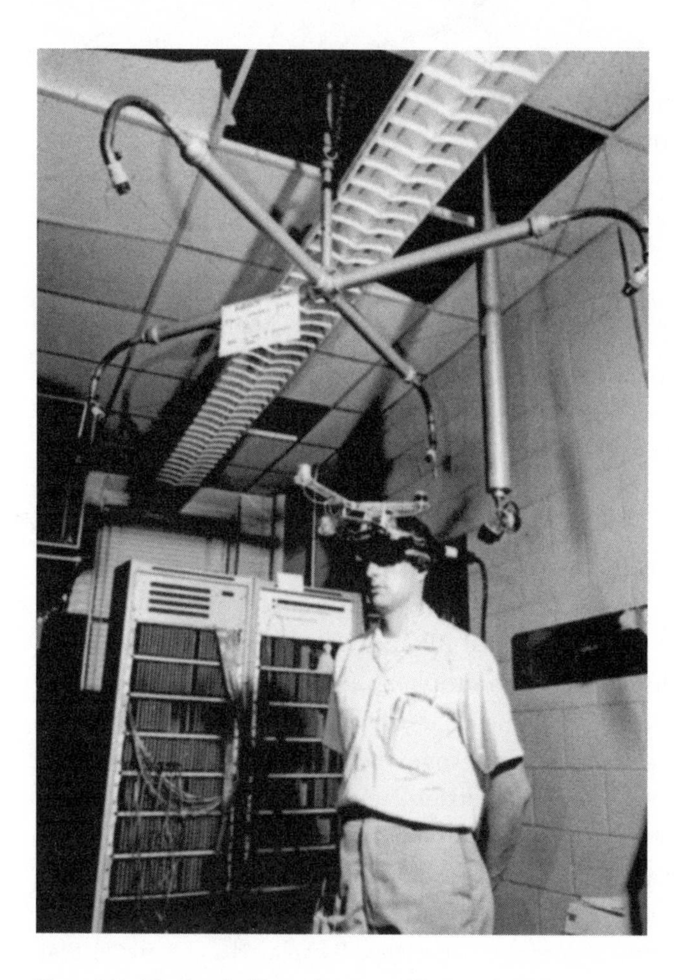

Figure 1.5 The Sword of Damocles, invented by Ivan Sutherland, advanced the concept of a head-mounted display. Source: Wikipedia Commons.

Various researchers continued to explore and advance the technology. In 1981, Steve Mann, then a high-school student, placed an 8-bit 6502 microprocessor—the same chip used in the Apple II personal computer—into a backpack and added photographic gear, including a head-mounted camera. This created a wearable computer that not only captured images of the physical environment but also superimposed computer-generated images onto the scene. A device called an EyeTap allowed a user to have one eye on the physical environment and another viewing the virtual environment. Mann later became a key member of the Wearable Computing Group at MIT Media Lab.

Over the next decade, the technologies surrounding XR advanced considerably. Researchers continued to develop more advanced digital technology that led to more sophisticated AR and VR systems and subsystems. Headsets began shrinking and morphing into goggles and eyeglasses, and designers and engineers began integrating an array of components into AR and VR systems. These included buttons, touchpads, speech recognition, gesture recognition, and other controls, including eye-tracking and brain–computer interfaces. In 1990, Boeing researcher Tom Caudell coined the term "augmented reality" to describe a specialized display that blends virtual graphics and physical reality.

By the end of the decade, augmented reality had made its debut on television. In 1998, a Sportvision broadcast of a National Football League game included a virtual yellow first down marker known as 1st and Ten. Two years later, Hirokazu Kato, a researcher at Nara Institute of Science and Technology in Japan, introduced ARToolkit, which uses video tracking to overlay 3D computer graphics onto a video camera. This open source system is still used today, including on web browsers. By the next decade, AR also began to appear in vehicles. Sports cars and higher-end luxury vehicles included systems that projected the speed of the vehicle onto the windshield. This made it easier for a driver to avoid looking down from the road. Today, AR technology appears in numerous products, from toys to cameras and smartphones to industrial machinery.

The term "virtual reality" also appeared in the common vernacular around this time. Jaron Lanier, a computer scientist who had worked at the gaming company Atari began promoting the concept in 1987. His startup firm, VPL Research, produced VR components that together represented the first commercially available product. It included gloves, audio systems, head-mounted displays, and real-time 3D rendering. Lanier also created a visual programming language used to control various components and combine them into a more complete VR experience. About the same time, computer artist Myron Krueger began experimenting with systems that combined video and audio projection in a personal space, and Douglas Engelbart, who is best known as the inventor of the computer mouse, began developing more advanced input devices and interfaces that served as the starting point for many of today's AR and VR systems.

Immersive Technology Emerges

The first truly immersive virtual space was introduced by a team of researchers at the Electronic Visualization Laboratory (EVL)—a cross-functional research lab at the University of Illinois at Chicago. In 1992, Carolina Cruz-Neira, Daniel J. Sandin, and Thomas A. DeFanti demonstrated the Cave Automatic Virtual Environment (CAVE).[20] It delivered a far more realistic VR experience, including a holodeck that allowed individuals to view their own bodies within the room. The system incorporated rear-projection screens, a downward projection system, and a bottom-projection system to create the illusion of reality within the physical space. A person standing in the CAVE wearing 3D glasses could view objects floating and moving through the room.

CAVE aimed to solve a basic challenge: early HMDs were bulky and presented significant practical limitations, including use for scientific and engineering applications. The first generation of the technology incorporated electromagnetic sensors to track motion and movements. Later versions tapped infrared technology. Motion capture software pulls data from sensors embedded in the glasses and the room to track movements. The video continually adjusts and adapts to the motion. The CAVE projection system ensures that the glasses are constantly synchronized so that the person views the correct images in each eye. The room is also equipped with 3D sound emanating from dozens of speakers. The result is an immersive virtual space that allows a user to manipulate and manage objects in 3D.

Early versions of CAVE were significant because they advanced XR far closer to today's portable and mobile systems. In fact, the CAVE concept caught on quickly. In 1994, the National Center for Supercomputing Applications (NCSA) developed a second-generation CAVE system so that researchers could explore the use of virtual reality in various fields, including architecture, education, engineering, gaming, mathematics, and information visualization. Using CAVE, an automotive designer could study the interior of a prototype vehicle and gain insight into how and where to position controls. An engineer might view the interior of a high-rise building before it is built, and a scientist might peer inside molecules or biological systems.

Today, many universities and private companies—from design and engineering firms to pharmaceutical firms—operate CAVE systems. These spaces are equipped with high definition projection systems that use state-of-the-art graphics to create lifelike effects. They also incorporate 5.1 surround sound, tracking sensors in walls and haptic interaction to deliver instant feedback. Because CAVE tracks head, eye, and body movements, a user can wave a wand to control virtual objects and move them around at will. It also means that a surgeon learning a new procedure will know instantly if he or she makes an incorrect incision. Over the years, CAVE has evolved into an entire platform with different cubes and configurations to suit different needs and purposes. A commercial offshoot of the project, Visbox, Inc., offers 12 basic configurations, as well the ability to completely customize the design of the space.[21]

The Military Recognizes the Power of Virtual Tech

One of the drivers of virtual reality and augmented reality was the US military. The goal of enhancing weapons and improving training is as old as war itself. As the 1960s unfolded, Bell Helicopter (now Textron Inc.) began experimenting with head-mounted 3D displays. In 1982, Thomas A. Furness III, who designed cockpits and instrumentation for the US Air Force, turned his attention to designing virtual-reality and augmented-reality interfaces that could be used in training systems. The Visually Coupled Airborne Systems Simulator overlaid data, graphics, maps, infrared, and radar imaginary in a virtual space in a head-mounted display that later became known as the Darth Vader helmet. The device included voice-actuated controls along with sensors that allowed the pilot to control the aircraft through voice commands, gestures, and eye movements.

Later, Furness designed a "super cockpit" that generated higher resolution graphics. In the UK, researchers embarked on similar projects. By 2012, the US Army had introduced the world's first immersive virtual training simulator, the Dismounted Soldier Training System.[22] The goal wasn't only to improve training. DSTS generated cost savings by reducing the need to shuffle troops around for training exercises. "[The Dismounted Soldier Training System] puts Soldiers in that environment," stated Col. Jay Peterson, assistant commandant of the Infantry School for the US Army in an August 2012 news release.[23] "They look into it and all of a sudden, they're in a village. There [are] civilians moving on the battlefield, and there [are] IEDs and vehicles moving. ... If utilized right, you can put a squad in that environment every day and give them one more twist," he stated.

Over the last two decades, branches of the US military—along with the Defense Advanced Research Projects Agency (DARPA), have devoted considerable attention to the research and development of AR and VR technologies. This has included head-mounted devices, more advanced control systems and wearable components.

During the 1940s and 1950s, considerable effort went into creating realistic flight simulators to train both military and commercial pilots. In 1954, the modern era of simulators arrived, when United Airlines spent $3 million for four systems. Today, the technology is mainstream, allowing pilots to practice and perfect procedures without risking a multimillion-dollar aircraft. More advanced simulators also train astronauts for space missions. Virtual reality has allowed the experience to become increasingly realistic.

Yet, the technology has steadily evolved beyond flight and training simulators and become a core component in ships, armed vehicles, and other systems. For instance, Battlefield Augmented Reality System (BARS), funded by the US Office of Naval Research, recognizes landmarks and aids in identifying out-of-sight team members so that troops can remain coordinated and avoid unintentional shootings.[24] BARS ties into an information database in real time and it can be updated on the fly.

Gaming and Entertainment Produce New Virtual Frontiers

It should come as no surprise that gaming drives many of the advances in computing and digital technology. What's more, it monetizes concepts and propels them into the business world. Accordingly, computer and video games featuring XR raced forward during the 1990s and 2000s. In 1991, Sega developed the Sega VR headset. It consisted of two small LCD screens and stereo headphones built into a head-mounted display. The HMD tracked eye and head movements. It could be used with games involving battle, racing, and flight simulation, among others.[25] However, the device was never released commercially because, according to then-CEO Tom Kalinske, it caused users to suffer from motion sickness and severe headaches. There were also concerns about injuries and repetitive use problems.

The first networked multiplayer VR system also appeared in 1991. It was named Virtuality. The technology, designed for video arcades, cost upward of $70,000. It was remarkable because it also introduced the idea of real-time interaction. Players could compete in the same space with near-zero latency. The project was the brainchild of Jonathon Waldern, who served as managing director for Virtuality. "It was a concept to us that was perfectly clear but to others we went to for financing it was crazy. They just couldn't imagine it," he stated in a later interview.[26] Nevertheless, Virtuality enjoyed modest popularity and companies including the likes of British Telecom purchased systems in order to experiment with telepresence and virtual reality.

By the 1990s, Atari, Nintendo, Sega, and other gaming and entertainment companies had begun experimenting in earnest with virtual reality. The film *The Lawnmower Man* introduced the concept of virtual reality to the masses. In the movie, a young Pierce Brosnan plays the role of a scientist who uses virtual-reality therapy to treat a mentally disabled patient. The original short story, written by author Stephen King, was inspired by VR pioneer Jaron Lanier. As the century drew to a close, another landmark film hit movie theaters: *The Matrix*. It featured people living in a dystopian virtual world. The film was a blockbuster success and imprinted the idea of virtual worlds on society.

Gaming consoles featuring virtual reality also began to appear—and often quickly disappear. Nintendo's Virtual Boy was released in Japan in July 1995 and the following month in the United States. The video-game console was the first to deliver stereoscope 3D graphics using an HMD. A year later, however, Nintendo pulled the plug on the project for a number of reasons, including high development costs and low ratings from users. The console did not reproduce a realistic color range—hues were mostly red and black—and those using the console had to endure unacceptable latency. In the end, fewer than two dozen games were produced for the platform. It sold only about 770,000 units worldwide.

Undeterred, engineers continued to focus on developing a viable VR gaming platform. Aided by increasingly powerful graphics chips that produced quantum leaps in video, consoles such as the PlayStation 2 and 3, Xbox, and Wii began using haptic interfaces, goggles, and new types of controllers. Yet, it wasn't until 2010 that modern VR began to take shape. The Oculus Rift, with a compact HMD, introduced more realistic organic light-emitting diode (OLED) stereoscopic images and a 90-degree field of vision. Over the years, the Oculus platform has continued to advance. In 2014, Facebook purchased Oculus from founder Palmer Luckey for $2 billion. The company has since built Oculus into a major commercial VR platform, and continues to introduce more advanced platforms, including the Quest, which it touts as" the world's first all-in-one gaming system built for virtual reality." At the same time, other companies have streamed into the VR marketplace. This includes Sony's Project Morpheus, a.k.a. PlayStation VR.

The Modern Era of AR and VR Arrives

Producing an ultrarealistic and ultrauseful XR experience requires more than hardware, software, and sensors. It demands more than incredible graphics and creative ideas.

It's essential to tie together disparate technologies and coordinate devices and data points. For augmented reality, this means managing real-time data streams through mobile devices and the cloud—and applying big-data analytics and other tools without any latency. For virtual reality, it's essential to design and build practical and lightweight systems that can be worn on the body. The Oculus Rift changed the VR equation by showcasing a lightweight platform that was both practical and viable.

The evolution of XR to smaller and more compact—yet still powerful—systems marches on. Over the last few years, wearable components and backpack systems have begun to emerge. Today hundreds of companies are developing and selling VR systems in different shapes and forms. Not surprisingly, prices continue to drop. Consider: the Oculus Rift, released in 2016, introduced a complete VR platform for $600. By 2018, a far more advanced Oculus Go sold for $199. Meanwhile, computer manufacturer HP has released a backpack system, the HP Z, that can be used not only as a conventional computer but also as a mobile virtual-reality platform. It weighs about 10.25 pounds and it comes with hot-swappable batteries and a mixed-reality head mounted display.

Augmented reality is also speeding forward. In recent years, automobile manufacturers have begun using AR for heads-up displays that show how fast the vehicle is moving. Google Glass, introduced in 2013, projected information on actual glasses and incorporated natural language commands, a touchpad, and internet connectivity so that a user could access a web page or view a weather forecast. The light-emitting diode (LED) display relied on a technique known as s-polarization to reflect light onto the lens. Google Glass also featured a built-in camera that could record events in the user's field of vision. In April 2013, Google released a version that cost $1,500. Although groups ranging from doctors to journalists began using the glasses, Google halted production of the consumer prototypes in 2015. It refocused its efforts on developing AR glasses for the business world.

Despite technical limitations and privacy concerns—including capturing audio and video in places many would deem "private" and inappropriate, such as a workplace or locker room—Google Glass took AR beyond the smartphone and into the form factor of lightweight glasses. Of course, the idea of projecting images and data on lenses isn't limited to Google. Dozens of other companies have developed eyewear that puts AR clearly into view. This includes Microsoft, Epson, and Garmin. Moreover, AR is continuing to expand beyond conventional screens, lenses, and uses. For instance, Apple has developed an augmented-reality windshield that can display map directions and a ccommodate video chats in autonomous vehicles.[27]

Although AR systems may use holographic projections to generate images on an LCD or other glass screens, they're steadily gaining other features, including audio, haptic, and laser interaction. What's more, smartphone apps that incorporate AR are increasingly common. These apps have both business and consumer appeal. For instance, they allow technicians to view data and specs while repairing a machine. Consumers can see what a new sofa might look like in a family room or what a garden will look like with roses rather than hedges. AR apps can also simplify travel and the stress of communicating in a foreign language. For example, Google Translate serves up real time translations for signs, menus, and documents. Simply hovering the phone above text creates a real time overlay in the desired language.

Some industry observers have deemed AR the new personal assistant, in much the way Apple's Siri, Microsoft's Cortana, Amazon's Alexa, and Google Assistant have redefined the way people interact with digital devices. At the center of this thinking is a basic but profound concept: the right combination of tools and technologies can result in dramatically fewer steps and far better results. They can also add new features and capabilities that weren't possible in the past. The fact that upward of 2.6 billion smartphones exist worldwide makes the technology even more appealing. Suddenly, it's possible to use AR *anywhere, anytime*.

In 2017, Apple raised the stakes by incorporating an AR development platform, ARKit, into its iPhone X

(Google later introduced its own kit, ARCore, for Android phones). One interesting and novel feature that Apple introduced was animated characters known as Animojis. The iPhone X uses facial recognition and a camera technology called TrueDepth to capture a person's image and embed it into the animated creature. Using 30,000 dots of infrared light to capture the features of a face, it's possible to mimic user expressions and motions in the Animoji. This includes smiles, smirks, frowns, laughs, raised eyebrows, and other gestures.

XR Gets Real

The appeal of AR and VR is indisputable. Humans live in a 3D world that serves up sights, sounds and other sensations on a 24 × 7 basis. It's how we define and experience our world. Although 2D television screens and computer displays have advanced remarkably, they deliver an experience that is nothing like the physical world. They cannot duplicate or recreate the physiology of vision, movement, sound, and touch. On the other hand, extended reality creates a more complete and immersive sensory experience that allow us to extend our consciousness into new realms and explore new places and things.

Jason Welsh, managing director for Accenture's Extended Reality Group in North America, believes that consumers and business are ready to embrace these new environments. "Over the next 10 years we will see an enormous shift in social behaviors," he says. Gaming will take on new dimensions and activities such as watching movies, and experiencing sports, concerts, and travel will become highly immersive. Virtual business meetings will assemble people from all over the world and businesses will use sensor data and feedback data from AR and VR to learn about customer behavior in deeper and broader ways. These technologies could even shape the future of space travel. NASA has introduced Mars 2030, a virtual-reality simulation where participants enter a virtual space and build communities on the red planet. The agency plans to use the data to plan actual missions.[28]

Of course, additional technology advances are required to produce AR apps and VR systems that slide the dial to mass use. For now, AR graphics continue to advance and, alas, smartphone apps and goggles don't always work as billed. There are other challenges, too. VR systems must become smaller and better integrated with the human body. Battery performance must improve. And persistent and ubiquitous network connections still aren't available everywhere. Further advances in microchips, software, and network designs are necessary to take XR performance to the next level.

Nevertheless, extended reality is taking shape—and reshaping society. London College of Communication began offering a master of arts degree in VR in the 2018–2019 academic year.[29] Another university, L'École de design Nantes-Atlantique in France, introduced a master's degree in virtual reality in 2010. Jaron Lanier, in a November 2017 *Wired* magazine story, offered a compelling perspective.[30] "Virtual reality is a future trajectory where people get better and better at communicating more and more things in more fantastic and aesthetic ways." Although extended reality is often considered a magical thing and is thought of as a pale comparison to the physical world, Lanier views things differently. "Virtual reality—which is sold as an illusion—is real for what it is."

2

THE MANY SHAPES AND FORMS OF AR AND VR

A Journey into a New Reality

At the most fundamental level, extended reality resides at the intersection of human and machine interaction (HMI). In truth, it represents a continuum that can incorporate a wide array of physical and virtual objects. This, understandably, leads to dramatically different results—and environments. In 1994, Paul Milgram, a professor at the University of Toronto in Canada, along with fellow researchers, introduced the idea that real objects and virtual objects could be juxtaposed to produce different forms, factors, and experiences.[1] These range from a totally real environment to an augmented-reality framework that becomes "augmented virtuality" and, in the end, a fully immersive virtual space.

Milgram's interpretation of the technology provides a good, if incomplete, starting point for navigating the space. Others have advanced the idea that physical reality resides at one end of the spectrum while digital reality lies at the other. Between these markers there are an almost unlimited number of combinations, recombinations, and mashups of AR, VR, and MR. In truth, extended reality is both simple and complex. On one hand, anything that augments or changes how we encounter objects and people in our natural state could be defined as an extension of reality. On the other hand, today's digital systems break new virtual ground by creating more immersive, convincing, and sensory-rich experiences. Computers bend, distort, and completely reshape the way we experience the world—and, along the way, how our brain processes information.

To be sure, augmented reality, virtual reality, and mixed reality are not monolithic terms or simplistic concepts. They encompass a wide variety of tools and

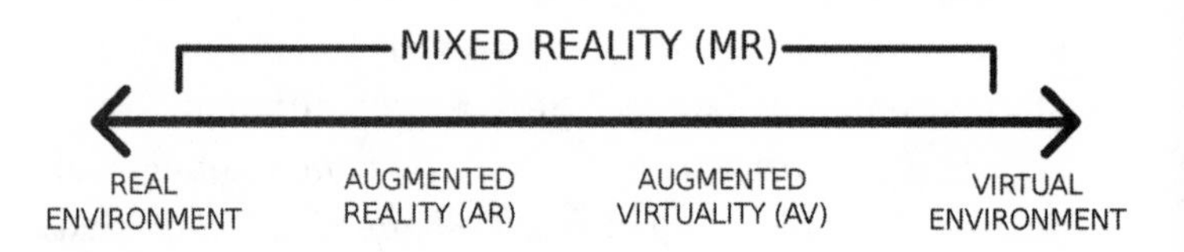

Figure 2.1 The continuum of physical and virtual objects.

technologies that touch all corners of the digital world. These include devices such as head-mounted displays, virtual retina displays, goggles, and glasses; binoculars and monoculars; haptic gloves; earbuds; and various other sensory feedback systems. Virtual environments may require high-definition photographic images and film, or graphics created by an artist or generated by a computer. Augmented- and mixed-reality environments might also include displays and visual overlays generated by a computing device and include audio cues and haptic feedback.

In most cases, it's the combination of technologies that makes XR systems so realistic—and so compelling. Designing, engineering, and integrating these systems into a completely seamless experience is at the heart of AR, VR, and MR. Yet it's no simple task to augment or replace physical reality. Tricking the human eye and mind into believing that the artificial environment is the same as a physical environment requires tightly integrated sensors, systems, and devices. Components such as motion detectors, accelerometers, microphones, cameras, skin sensors, and other types of monitoring and tracking systems are essential. What's more, the software that drives XR requires complex algorithms and enormous processing power, increasingly in the cloud, to generate convincing virtual environments.

XR Technology: The Basics

Although visual displays and applications are a primary focus for AR, VR, and MR, the presentation framework and platforms used to generate these technologies are broad. Depending on how designers and engineers assemble various components, it's possible to achieve very different tools, form factors, and virtual spaces. Let's take a look at some of the forms and form factors for extended reality.

Augmented Reality on Smartphones

Augmented reality blurs the distinction between what's part of the physical world and what is digitally generated. Computers may add graphics, text, images, and video to an existing scene in real time. They may also add sounds, including speech, to an app. In a basic sense, when someone uses Apple Maps or Google Maps and hears voice prompts about how and where to proceed, he or she experiences a type of AR. Augmented reality can also incorporate haptic feedback, such as when a phone provides a tactile sensation to confirm that a user has entered a command or performed a specific gesture.

It's no news flash that smartphones have accumulated functions and features that previously resided in multiple devices and objects, including cameras, audio players, recording devices, book readers, GPS systems, and maps.

Augmented reality blurs the distinction between what's part of the physical world and what's digitally generated. Computers may add graphics, text, images, and video to an existing scene in real time. They can also add sounds, including speech, to an app.

The combination of these tools and technologies—and the way they interact in groups or together—frequently creates a sum that's greater than all of the individual components. In fact, many features and capabilities simply didn't exist a few years ago. For example, smartphones and digital photo software allow users to modify images and add effects and filters that instantly change the way a person or scene looks.

Of course, adding advanced augmented-reality features to smartphones requires additional layers of technology. Not only is it necessary to tie together hardware components—such as a camera and microphone—there's a need for powerful onboard processing to capture and render images and data, along with specialized algorithms to manage it. What's more, some of this data must be sent to the cloud for processing. The result—such as a data overlay or a graphic representation—must be visible in real time. Even the slightest lag in performance can detract from the AR experience. In a worst-case scenario, it can render the feature or app useless.

Consider: Google Translate includes a feature that allows a smartphone user to hover the camera over a sign, menu, book, or document and view a live translation. Banking apps superimpose a frame around a paper check so that a user can hover the smartphone camera over it and capture the image for a deposit. Retailing giant IKEA offers an app that lets a user select a piece of furniture from a smartphone screen, hold the camera up to a room, and visualize the desk or table in a physical space. Apps such as Facebook and Snapchat allow users to draw on top of photos or videos and manipulate them using a variety of special effects that include flowers, animals, and graphic designs.

In the business world, AR unleashes an array of capabilities. It might allow a real estate agent to view information about appliances or the dimensions of a room while walking through a house with a potential buyer. It might help a contractor or technician view specs for plumbing, electrical, and other systems while a structure or facility is under construction. And it introduces opportunities for groups to share documents or view data in new ways, including at physical and virtual meetings. Already, an app called Augment lets sales reps demonstrate in 3D what a grocery store display will look like to a store manager.[2]

AR development kits, such as Apple's ARKit and Google's ARCore, have introduced new and remarkable capabilities. These platforms also continue to become more powerful, partly driven by microchips and processing platforms that are optimized for AR and VR. They use sophisticated motion tracking to observe objects along with their relationship to a space. This makes it possible to detect horizontal and vertical surfaces in the physical world to understand how to position and move objects in an AR or VR environment. They also observe light within a physical environment so that virtual objects appear more realistic.

AR on Glasses, Lenses, Goggles, and Heads-Up Displays

The first encounter most people had with augmented reality occurred in July 2016. That summer Pokémon Go emerged as a sensation. People of all ages began chasing down cartoon characters that appeared on their phones as they held up the cameras in the physical world. This allowed them to embark on the digital treasure hunt. At the very least, it got a lot of children—and some adults—up and moving. But it also showed what's possible within the realm of AR. Today, a growing array of AR activities and games exist. These include other apps that use geolocation to send users on treasure hunts, shooting activities, and sports games that approximate golf and basketball.

At the center of all this is the smartphone. It allows users to glance at the screen or listen to voice prompts or other audio cues while sitting, standing, or moving. These capabilities completely rewire digital interaction by putting powerful tools at a person's fingertips, *anywhere* and *anytime*. Therefore, smartphones are ideal for a vast array of AR applications and services. Holding a phone in certain situations can range from inconvenient to dangerous, however—particularly if an individual is engaged in an activity like driving an automobile or standing on a ladder with a screwdriver or hammer in the other hand. This makes it desirable, and sometimes crucial, to display data in a hands-free mode.

One widely used form factor for hands-free augmented reality is a heads-up display (HUD). The technology projects crucial data—anything from the speed a car is traveling to the horizon line or target for a military fighter jet—onto a glass windshield or display screen. This helps a driver or pilot avoid looking down at instruments or gauges for even a split second. The technology made its debut in military and commercial aircraft in the 1970s and it has since appeared in vehicles, industrial machines, and even toys. Not surprisingly, heads-up displays continue to evolve. Newer systems tie into onboard sensors, including GPS systems and infrared cameras, to provide enhanced vision in poor weather conditions and in the dark.

While the technology continues to shrink, the capabilities grow. Today, a pair of lightweight glasses or goggles can project images and data over a view of the physical landscape or space. In the case of a mountain climber, this might include showing the altitude, weather conditions, and distances. For an insurance adjuster reviewing fire damage, an AR app might capture images of the damage, transmit them to a cloud-based server, and offer a client an immediate analysis of costs and reimbursements. These same AR systems transform a variety of 2D activities—from board games to amusement park visits and sporting events—into richer and more intriguing environments.

The possibilities don't stop there. With GPS and mapping, a pair of goggles can help a person arrive at a destination by displaying arrows or a breadcrumb trail along a route, along with augmented street names or landmarks as they come into view. The AR app might also include whatever relevant data the user desires, including subway schedules, nearby restaurants or coffee houses, and even ratings for businesses along a route. Integrated speech recognition, video capture and audio instructions allow an individual to accomplish tasks without removing a smartphone from a pocket or purse.

Google Glass is perhaps the most widely known example of AR glasses. The product became publicly available in May 2014 for $1,500 but it was discontinued in January 2015 as a result of privacy concerns and a public backlash (users not so affectionately became known as "glassholes"). Google has since introduced Google Glass Enterprise Edition, which focuses on specialized business uses. The AR glasses include a graphical development toolkit for creating applications along with numerous APIs that connect to services and other apps. Limited processing power, though, has so far limited their use cases.

Microsoft's HoloLens is also changing the AR picture. The $3,000 goggles—designed primarily for business

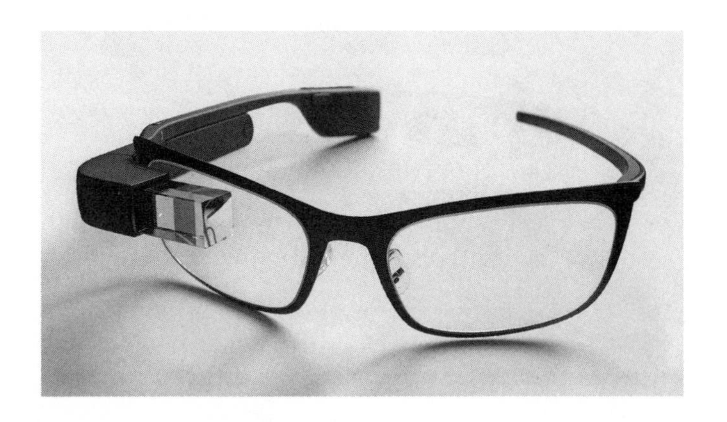

Figure 2.2 Google Glass with Frame. Source: Google.

users—offer a platform for augmented reality and mixed reality development. The system runs Windows apps by rendering 2D and 3D holograms on the glass surface. This allows a user to view a blank wall as a monitor and manipulate the environment with hand gestures. A leading design and architectural firm, Genser, has used the HoloLens to redesign its Los Angeles headquarters.[3] The system, with specialized software called SketchUp Viewer, projected 3D mixed-reality images and holograms directly into the eyes of designers and engineers. They could preview, view, and understand components and spaces—including size, scale, and various perspectives—through the visuals.

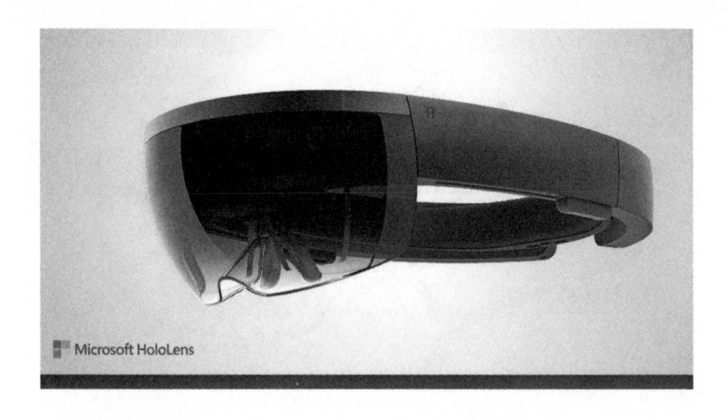

Figure 2.3 The Microsoft HoloLens. Source: Microsoft.

Other companies are using AR glasses and goggles to completely revamp the way they handle tasks in places as diverse as office spaces and industrial facilities. For example, DAQRI Smart glasses allow construction workers, engineers in plants, and oil rig workers to view live digital data about machinery or the environment they're working in.[4] AR manufacturers are also exploring how to embed sensors and systems into clothing and other wearables, so that a smartphone or other computing device won't be required to power the AR or MR display. Facebook CEO Mark Zuckerberg has stated that AR could eventually replace all screens, from smartphones to TVs. What's more, these systems might eventually be controlled directly by the human mind.[5]

Virtual Reality on the Body

Over the last few years, a steady stream of news reports has trumpeted the possibilities and opportunities presented by virtual reality. Consumer headgear and other VR systems from companies such as Oculus, Sony, HTC, Lenovo, and Samsung have become increasingly popular. Sony reported in August 2018 that it had sold more than 3 million PlayStation VR headsets and 21.9 million "games and experiences."[6] These include titles such as *Batman: Arkham VR* and *Resident Evil 7 Biohazard*.

Fueling adoption is the fact that the technology is advancing at a furious clip while prices are plummeting. Many of VR systems, which initially sold in the $800 to $1,000 range, now cost less than $200. At the center of this revolution are smaller, faster, and more powerful central processing units (CPUs) and graphics processing units (GPUs). The latter chips, produced by Nvidia, Qualcomm, AMD, Intel, and Asus, dramatically improve VR rendering speeds while generating more realistic graphics. They have replaced slow, buggy, and distorted graphics with ultrarealistic visuals that please the eye and trick the mind. This makes it possible to venture into exotic fictional worlds as well as extensions of the physical world without pause.

Consequently, numerous companies—including digital marketing and media agencies—are wading into virtual reality through HMDs that connect to a computer or smartphone or operate completely untethered.

They're introducing an array of apps that stretch—if not redefine—the concept of visiting a physical space or place. Likewise, entertainment and gaming companies are advancing VR by creating environments that incorporate high-definition graphics or actual photographic images. They may also incorporate sound and haptic feedback to create a more realistic sensory experience.

The range of devices and price points is impressive. Google has introduced an entry-level VR with a platform called Google Cardboard, which it introduced in 2014. "Experience virtual reality in a simple, fun, and affordable way," a Google tagline notes. Using this system is simple. A user simply inserts a smartphone into a foldable cardboard holder that costs as little as $6 and, suddenly, news, games, travel, and more gain 3D characteristics. Google Cardboard uses a software development kit (SDK) that taps stereoscopic techniques. By early 2017, Google had shipped more than 10 million Cardboard viewers and tallied over 160 million Cardboard app downloads.

At the other end of the spectrum: a growing array of high-end systems for both consumers and businesses. For instance, Magic Leap has developed a $2,295 ultralight mixed-reality headset called One Creator Edition that incorporates advanced eye-tracking, a controller that approximates human motion, super-sharp visuals, and software that renders ultrarealistic scenes. The HMD looks like a space-age set of goggles rather than a bulky and cumbersome headset that requires straps. The wireless system is powered by a "lightpack" that clips into a pocket.

Virtual-Reality Spaces

Although Google Cardboard makes it possible to hold VR in the palm of one's hand, it falls squarely into the fun and games category. By contrast, the Cave Automatic Virtual Environment (CAVE) environment, invented by researchers at the University of Illinois at Chicago, has evolved into a virtual-reality theater that's widely used by engineers, scientists, and major corporations.[7] The environment is particularly popular among industries engaged in collaborative planning and construction—although CAVE has been used to better understand everything from how to better design manufacturing spaces to how to land a jet aircraft in turbulent conditions. CAVE also allow students, faculty, and others to explore data representations in 3D and conduct research in new and interesting ways.

At some point in the future, the immersive physical environment—which supports a robust virtual-reality framework—won't be necessary. Advances in technology will replace the need for sophisticated hardware with wearable devices along with software and algorithms. Yet, for now, CAVE is a core tool for organizations using VR. What's more, the newest versions of CAVE integrate digital technologies and provide features and capabilities that wearable systems do not.

For example, at Louisiana State University (LSU), a new Building Information Model Lab (BIM) taps 11 PCs along with sensors for movement, temperature, and humidity, to create a highly interactive environment with 300-degree visuals.[8] The 2,400-square-foot room contains 44 OLED display screens—each with a diagonal measurement of 55 inches—along with state-of-the-art audio and haptic feedback systems. It bypasses headsets and instead

Figure 2.4 The CAVE virtual-reality environment with an omnidirectional treadmill. Source: Wikipedia Commons.

focuses on creating an immersive environment through the flat-screen displays.

"The CAVE provides a certain amount of immersion, but the viewpoints are very different. It gives you a sense of being in a certain environment," stated Yimin Zhu, a professor in the Department of Construction Management at LSU. "There are two purposes for the CAVE. One is data visualization. The CAVE is a visualization platform that provides us with stereoscopic views for many participants at the same time. The second purpose is for teaching. We do have other means to visualize 3-D or virtual models but many of them are either on the desktop or use the head mount, which is more immersed but individual."[9]

The CAVE is convincing and compelling because it generates a lifelike visual display—along with sounds and other feedback. It's also highly flexible. In some cases, it may incorporate glasses or goggles that sync with the environment and produce a 3D image but, as the LSU space demonstrates, it can also incorporate a wall of flat screen displays to simulate images and motion directly with the human eye or through active shutter 3D glasses. The latter use liquid-crystal technology to alternatively display and block images to trick the brain into seeing a single 3D image with minimal flicker. The same active shutter 3D technique is used for viewing 3D films in theaters.

Still another thing that makes CAVE appealing is the availability of free or open source software libraries that allow users to build scenes and manage events within the VR environment. These include tools such as OpenSG, OpenSceneGraph, and OpenGL Performer. For example, Open SG, which is written in C++ programming language, runs on Windows, OS X, Linux, and Solaris operating systems. It uses sophisticated multithreading to manage computing memory and resources. This aids in creating scenes where graphics display in real-time and sync with user movements and other activities.

Haptics Takes Hold

Because humans experience the world through five senses, the ability to view and hear virtual activities is only part of the story. A sense of touch and feel—what scientists refer to as haptic feedback—is also a critical element in augmented-reality, virtual-reality, and mixed-reality environments. Mechanically generated sensations can mimic what an actual physical object or thing feels like. Something in the virtual environment—it could be petting a dog or opening a door to a house—can assume the feeling of the actual object or act by providing the appropriate level of vibration and pressure.

Haptic feedback isn't a new concept. Many smartphones already incorporate it into functions. It's also used in fields as diverse as robotics and surgery—and it's commonly found in high-end flight simulators, surgical simulators, and industrial systems. The technology, which has been around since the 1970s, has also played an integral role in creating more realistic gaming consoles and delivering feedback on other devices. Tactile feedback may create a sensation of a string pulling lightly against the finger or that of a person having pushed a mechanical button. It can notify a smartphone user that he or she is checking e-mail or has selected an object in an app. In games, haptic feedback frequently offers cues about navigating the environment or achieving the desired result.

Not surprisingly, incorporating haptic feedback into smartphones, game controllers, and other devices is quite different than producing a realistic sense of touch in a virtual-reality world. VR systems require more advanced sensors for detecting user movements and more sophisticated software that can map motion, manage the complexities of various objects within the space, and integrate visuals and touch seamlessly. It's not good enough to deliver a general approximation of feel; the coordination and resulting experience must be precise. While researchers and companies have experimented with haptics and some have developed gloves that deliver basic sensations, such as vibrations, the ability to feel objects realistically has proved elusive.

More advanced haptics is on the virtual horizon, however. For instance, HaptX, a Seattle firm, has developed a sophisticated glove system called AxonVR. The company claims that the technology is sensitive enough to let the user detect an object the size of a grain of rice. A person wearing one of its glove systems can reach into a computer-generated virtual world and engage in a highly interactive experience with animals, plants, and other objects. For example, it's possible to flick a cloud and trigger a rainstorm—and then feel the raindrops falling onto one's hand. It's also possible to feel shafts of wheat growing in a field and the tickle of a friendly spider crawling across one's hand. The HaptX virtual world is comprised of cute animated objects and animals, but it could just as easily incorporate the ultrarealistic feelings of a surgical procedure or a space battle.

The ramifications shouldn't be lost on anyone. Research in the field is expanding at a rapid rate and an understanding of how to create a realistic touch experience is growing. "Haptics systems are valuable, if not essential, for almost every type of VR interaction," states Jake Rubin, founder and CEO for HaptX. "As VR and AR advance and the need for more realistic and immersive systems grows, physical touch and feedback are critical." The challenge—it's the same one that exists for head-mounted displays—is how to miniaturize equipment and make it affordable across a spectrum of industries and uses. Likewise, haptics systems integrated with VR must be flexible and adaptable for a wide range of uses—from the operating room to a construction site.

The end goal is to introduce VR and AR environments that create the illusion of reality across the sensory spectrum. This would allow a medical student or resident in training to practice a surgical procedure 50 or 100 times in a virtual environment—before transitioning to real-world surgery. At that point, the pressure and stress of operating on a real person would presumably subside. The task would seem completely natural. The same type of ultrarealistic virtual-reality simulation would allow a prospective pilot to practice takeoffs and landings until he or she has the technique mastered in the virtual space. Again, by the time the student sits at the controls of an actual airplane, operating the aircraft is second nature.

Gloves are one part of the evolution of haptics. Over time, they will continue to become more compact and powerful. David Parisa, an associate professor of emerging media at College of Charleston in South Carolina, believes that the endgame is a "master device"—a full-body haptic interface capable of replicating a whole range of haptic sensations.[10] Rubin says that omnidirectional treadmills and other specialized machines combined with lightweight body suites equipped with sensors could introduce a level of virtual interaction that would be profound. This would make it possible to run along a trail in Tierra del Fuego, Argentina, swim among fish in the Great Barrier Reef of Australia, and even experience simulated sexual activity. To be sure, the boundaries of extended reality are limited only by the ability of humans to invent them.

Venturing into an Extended World

The sum of virtual tools and technologies could usher in a very different world in the years ahead. Suddenly, it would be possible to walk through a virtual store, peer at various products, see how they actually work, and toss them into a virtual shopping cart. It would be possible to explore places around the world—famous buildings, caves, waterfalls, and landmarks—without enduring a long journey and suffering from jet lag. The key question, of course, is whether a virtual experience can replace the physical and visceral experience of shopping or travel. It's more likely that XR technologies will play a niche or supplemental role. Likewise, XR is unlikely to replace the web. It may, however, force sites to change, morph, and evolve into new and entirely different forms, including offering 3D experiences.

There are bigger questions about whether the technology will create a safer world or usher in a surveillance

Virtual tools and technologies could usher in a very different world in the years ahead. Suddenly, it would be possible to walk through a virtual store, peer at various products, see how they actually work, and toss them into a virtual shopping cart, and to explore places around the world without enduring a long journey and suffering from jet lag.

state. In China, for example, law enforcement officials now rely on facial-recognition glasses to spot and apprehend suspected criminals. The AR system uses lightweight eyeglasses that record images, feed them to the cloud, and, in real-time, provide matches for those engaged in illicit activity. The technology can parse through a database of 10,000 suspects in 100 milliseconds or less. In early 2018, police at Zhengzhou East Railway Station nabbed seven fugitives connected to major criminal cases and identified 26 others attempting to travel using fake IDs, according to *People's Daily*, China's official state newspaper.[11]

Nearly half a world away at San Francisco State University, the kinesiology department is exploring virtual reality as a way to stay fit.[12] A team of researchers, led by Professor Maryalice Kern, have developed virtual-reality games to promote exercise and physical fitness. "Nearly all virtual reality games involve some form of movement," Kern noted. "Some are as simple as turning your head from left to right, but others require very vigorous movement, like dancing. What we wanted to know was: how much energy do people expend while playing VR games, and can it really be considered exercise? ... These highly customizable and engaging tools may be very important to the future of how people stay healthy."[13]

The research group collected heart rate data and oxygen consumption data from those involved with virtual activities such as archery and boxing. They studied the metabolic data and ultimately found that virtual-reality exercise delivers real-world benefits. For instance, those engaged in virtual boxing burned 13 to 18 calories per minute. This compares to burning approximately 15 calories per minute doing the real thing.

XR will also change the way we absorb news and information. Although CNN, the *New York Times,* and other news networks are already experimenting with augmented and virtual reality, the future of news delivery will look very different than today's flat-screen televisions. Suddenly, with VR technology, it's possible to experience at least some of the sensations and realism of a hurricane, an earthquake, or a nuclear blast without succumbing to the destructive forces of these events. Just as illustrations changed newspapers and books in the seventeenth century, and then photos introduced greater realism in the nineteenth century, extended reality offers insight and realism far beyond what words and photos can provide.

In March 2018, the *New York Times* presented a story called "Augmented Reality: David Bowie in Three Dimensions."[14] The article and graphics focused on the legendary performer's costumes. Using AR, the story was transformed into 3D. Readers using a web browser on a personal computer viewed Bowie's costumes in a 360-degree view. As a user scrolled up or down across the graphic of the costume it would spin around. But the graphics took on an entirely different dimension on a smartphone or tablet that supported AR. On an iPhone X, for example, a user could float the costume in the physical room he or she was in and walk around it—while at the same time viewing the actual room along with all of the people and objects in it.

The *New York Times* provided insights into why it is using AR in a February 2018 post.[15] "If photography freed journalists to visually capture important moments, and video allowed us to record sight, sound and motion, then our augmented reality feature goes a step further, making flat images three-dimensional." The same AR features allow the newspaper to "see an athlete suspended mid-air as if she were floating in your living room. The camera can become a window into a world enhanced with digital information—adding a piece of sculpture to your bedroom or a car to your driveway. Neither actually there, but appearing to be and believably so."

The technology also has changed the way the military approaches combat situations and soldiers fight wars. Already, AR and VR have been used for flight simulation, battlefield simulation, medic training, and virtual boot camp. Psychologists have also tapped XR for preparing and "hardening" troops as well as treating soldiers afflicted with post-traumatic stress syndrome (PTSD). These technologies deliver cost savings but also provide clear improvements in care. For instance, a virtual-reality program called BRAVEMIND, created by a group of researchers at the University of Southern California (USC), allows individuals with PTSD to explore traumatic experiences while clinicians tweak conditions and parameters to help patients confront and process challenging emotional memories in order to work through their trauma.[16]

The project was originally funded by the US Office of Naval Research and headed by Albert "Skip" Rizzo, a psychologist and director for medical virtual reality at USC's Institute for Creative Technologies. It demonstrated in an initial pilot trial that 16 of 20 members of the military (19 men and one woman) showed improvements as a result of the virtual therapy.[17] Others have begun to use VR to aid amputees, stroke victims, and others. What makes these systems so powerful—and the reason the technology will likely revolutionize psychology, medicine, and many other fields—is that XR introduces immersive, interactive, and multisensory environments. A researcher or doctor can not only control the environment and receive instant feedback, but also adapt and adjust the VR, AR, or MR framework to respond to aggregate or individual data collected by sensors. This feedback loop allows care providers to home in on issues and address them in a way that wasn't possible in the past.

Yet, at the other end of the spectrum, VR gaming continues to take users to worlds that were once unimaginable. In 2016, Sony's Social VR introduced the idea of players using a beach ball to bounce into the sky and view trees, hills, oceans, and people from above. Reentry simulated the feeling of skydiving. Another game, *Waltz of the Wizard*, allows participants to feel as though they are a wizard who collects magical spheres and drops them into a huge cauldron. By doing this, the participant gains special powers, such as the ability to shoot off lightning bolts. Another interesting example is the film *Allumette*, which offers a virtual experience with an airship that floats through clouds. By physically walking, a participant changes perspective on various objects in the virtual space. Those stepping into the VR environment stated that they had to stand during the experience because the realism creates a sense of discomfort.[18]

The common denominator in this emerging world of XR is that augmented, virtual, and mixed realities alter perception and, along with it, actual reality. In a world where the lines between the physical world and digital world already blur, it introduces new and sometimes better ways to do things. Yet, all of this isn't possible without a sophisticated combination of digital technologies and software that can orchestrate the virtual experience. How all of this technology works ultimately defines what end users see and do. It's the foundation for the coming virtual revolution.

3

THE TECHNOLOGY BEHIND THE REALITY

Beyond the Real World

The technologies that collectively comprise AR, VR, and MR are evolving at a rapid rate. As microchips become more powerful, components shrink in size, and researchers develop far more sophisticated software, extended reality is taking shape. Yet, there's no single path toward augmented and virtual technologies. Different industries and use cases demand entirely different tools and components. A fully immersive gaming experience is markedly different than a system—or the headgear—a military aircraft pilot would use while flying a jet or an engineer would need to view a virtual building. An app on a smartphone is considerably different than augmented-reality eyeglasses a firefighter might rely on.

System design and engineering are critical to usability. Yet, the technical and practical challenges associated with creating a realistic virtual environment are formidable. This is particularly true for complex virtual-reality and mixed-reality environments. While it's apparent that a person entering a virtual space will require a realistic visual experience—through pictures, video or graphics—it's also necessary to tie in sensations, sounds, and other stimuli, including possibly scent, in order to convince the human brain that it's all real. Connecting and coordinating digital technologies are at the center of producing a convincing and meaningful virtual experience.

Consider: a person, say an architect, decides to walk through a virtual building and see what rooms and spaces look like before finalizing a set of designs and blueprints. This may include viewing walls, ceiling, flooring, architectural and design elements, and even layers of hidden infrastructure such as wiring, lighting, or plumbing. The problem is that walking through an actual room with a head-mounted display could mean tripping over physical objects, stumbling and falling, or crashing into walls or girders. Sitting down and bending into spaces introduces other challenges. All of this points to a need for a convincing virtual environment that's coordinated with the existing physical environment.

Similarly, augmented reality requires an understanding of physical objects and spaces, including the relationships between and among them. This might include words on a sign or menu, or the dimensions or characteristics of a room. In order to obtain this information, the AR system may require a camera or other type of sensor to measure a room's size or shape or discern the color or texture of the walls. If this system does not work correctly—say the AR app and the underlying algorithm can't determine the physical parameters and translate all the data points into physical representations—a piece of furniture will appear out of perspective or the color of the room won't look right.

The result is a poorly performing app or a useless experience. Grossly distorted rooms or furniture floating in the air will likely lead, at the very least, to a person abandoning the app. The consequences can be far worse, possibly even dangerous. An AR system that renders images or data incorrectly could result in an injury or death—particularly if the application is used in dangerous work environments, for policing, or in combat situations. What makes XR so challenging from a design and engineering perspective is that even the tiniest glitch or flaw in hardware, software, user interface, or network performance can torpedo the entire experience. Suddenly, the environment morphs from convincing to implausible.

How Augmented Technology Works

What makes augmented reality so enticing is that it allows humans to step outside the physical world—while remaining in the physical world. The visual piece of the AR puzzle is particularly important. Superimposing graphics, images, and text onto a smartphone screen or a pair of glasses is a complex multistep process. This varies, of course, depending on what the creator of an AR app or feature is attempting to do.

A starting point for developing an augmented-reality app is for a digital artist or graphic designer to create a 3D model, or for an app developer to incorporate a model from an existing 3D library from a private company or an open-source group. Specialized software such as 3DS Max, Blender, Cinema 4D, Maya, Revit, or SketchUp deliver tools that designers use to create visuals within an AR app. A platform like Sketchfab lets designers discover, share, and purchase content and components to create apps. An artist typically begins with a rough sketch or wireframe and, using these tools, refines the drawing until a 3D model takes shape.

Once a concept drawing exists, the artist transforms the drawing into an actual model. This animation process may include adding colors, shapes, textures, human features, or the behavior or physics of how an object or device operates in the physical world. In some cases, the goal may be to create a character that appears to have lifelike features. If the goal is to display an actual object—say David Bowie's "Ashes to Ashes" costume from 1980 (which appears in the *New York Times* app mentioned in chapter 2)—there's a need to capture an image or video of an item and convert it through the use of 3D projection mapping software. This transforms the 2D structure into a 3D representation.

The next step is to create the actual augmented-reality experience. After software transforms the mathematical construct into polygons—the fundamental geometry that underlies a 3D visual model—it's possible to add movement or give the person, creature, or object real-world features within the AR app. Yet, there's more to this task than simply creating an object of interest. When a person uses an app or AR glasses, the object must be visible from different angles and perspectives. Too few polygons and a virtual object won't look or feel real. Too high of a polygon count and the AR object may not render correctly.

The final step is to finalize the AR features using a software developer kit, such as Apple's ARKit, Google's ARCore, or an independent platform like Vuforia. These tools aid in building an app and embedding AR features. At this stage of the development process, app creators add things like motion tracking and depth perception—as well as area-learning features that integrate the virtual object with the actual physical environment. This helps the smartphone camera or eyeglasses capture and display the object correctly. The seamless blending of the physical and virtual objects is what makes an AR scene appear convincing and real. Without using these tools, the display device will not orient and coordinate all the elements. Motion may be off register and objects may appear distorted.

In fact, the ability to orient and coordinate physical and virtual objects—and manipulate the latter—is at the center of a successful AR app. Some augmented-reality SDKs, such as the Vuforia platform, include features that automatically recognize objects and compare them to existing 3D models built into the software framework. By estimating camera position and the overall orientation of objects it's possible to introduce a more realistic experience. Motion tracking software that operates in the app—simultaneous localization and mapping (SLAM)—creates the necessary depth tracking and pattern tracking.

SLAM, which is also used for autonomous driving and robotics systems, relies on algorithms and sensor data from the device to generate a 3D map of the space. It determines where and how to display the AR object on the smartphone or AR glasses. SLAM also uses global lighting models to render images with the right colors and tones. All of this information about registered objects is stored in a database that ultimately determines the visual outcome in the VR or AR environment. Then, using a rendering engine such as Unity or Unreal Engine, the phone or glasses generate the visual images. The iPhone X, for example, incorporates a TrueDepth sensor that projects 30,000 invisible infrared dots on an object. This allows sensors in the device to map a physical space and generate an accurate representation of a virtual object or an environment.

The final effect is the ability to render an object or scene in a way that appears photorealistic. A larger object typically means that the object is closer, and a smaller pattern implies that it is further away. The distance between objects helps determine the overall perspective of the scene—as well as how objects rotate and move in the AR space. The embedded algorithm—and the device's ability to use microelectromechanical systems (MEMS) such as accelerometers, proximity sensors, and gyroscopes—determines how the final scene appears on the phone or on a pair of glasses.

All in the Display

Although modern augmented-reality features and apps began to appear on smartphones as early as 2010, advances in electronics have transformed the space in recent years. For one thing, cameras in smartphones have improved to the point where they capture images in high definition. For another, today's OLED screens deliver high contrast ratios at an ultrahigh resolution. All of this allows colors and objects—including computer-generated images—to appear more real. SDKs and rendering platforms such as Unity and Unreal Engine have also continued to evolve.

Of course, form factors matter. A smartphone display offers the advantage of portability. The device is easy to tuck into a pocket or purse and it's convenient to use in a wide array of situations. The screen is available instantly and it is able to adapt and adjust to lighting conditions by dimming and brightening automatically. Yet a phone screen isn't as convenient as a pair of glasses or goggles, which reduce or eliminate the need for hands-on interaction with the device—particularly while using physical objects. For technicians, engineers, emergency responders, and many others, including consumers attempting to fix things—say replacing a light switch or sprinkler head—holding a phone to view written instructions or a video can prove challenging, if not impossible.

Indeed, AR glasses and goggles are all about going hands free. Google Glass is perhaps the highest-profile example of AR optics, but a growing number of firms are developing products for consumers and businesses. The list includes Epson, Sony, Microsoft, Nokia, and Apple. Prototypes and commercially available systems use several different technologies to project AR imagery onto lenses and connect to a smartphone or the internet. Form factors range from traditional eyeglasses to head-mounted displays, virtual retina displays (VRDs) that use low-powered LED light and special optics to project an image directly into the human eye without the use of a screen, and the heads-up displays used in vehicles and aircraft. HUDs use a laser, or a mirror that projects or reflects light onto the glass surface.

A key consideration in AR is field of view (FOV). Many of today's systems offer 50-degree or lower FOVs. This produces a scene that is best described as adequate but hardly compelling. Essentially, limited FOV restricts the field of vision and forces users to view objects within a narrower scope, which isn't natural. In order to achieve a superior view, a system requires an FOV of 180 to 220 degrees. A limited view also has potential repercussions for technicians and others who require a broader view of the space they are in. If the AR display obstructs their view they are more prone to make mistakes or cause an accident.

Display technology continues to advance. Researchers at the University of Arizona have begun to experiment with holographic optics.[1] This could produce a larger eye box—the display system that creates the image—and thus improve the clarity and viewing angles for data displayed on a screen or windshield. At the same time, researchers are looking for new and innovative ways to improve these systems. For instance, a Honda Motor Company patent granted in 2015 proposed to improve safety through an augmented-reality HUD that maps a forward view of a pedestrian along with a crosswalk path.[2] The system could flash a warning or apply the brakes automatically if it detects a potential hazard.

All Eyes on Wearable AR

Glasses, goggles, or any other device designed to project images onto the human field of vision operate on a basic principle: they redirect light into the eye. This includes the natural light generated from the physical world as well as the artificial digital images created by a computer. The latter typically takes the form of LEDs and OLEDs. The optical device must combine natural and artificial light in order to create augmented reality. The system uses what is referred to as a "combiner" to meld these light sources into a single field of vision. The combiner achieves this result by behaving like a partial mirror. It selectively filters light from each source to produce an optimal augmented image.

Wearable AR optical systems come in two basic forms: HMD optical combiners and waveguide combiners.[3] Each offers advantages and disadvantages involving shape, feel, aesthetics, weight, image quality, and resolution. Optical combiners rely on two different methods to generate images. The first is *polarized beam combiner* systems such as Google Glass and smart glasses from Epson. These glasses are typically lightweight and affordable, but they are not capable of generating the best possible image. The split-beam approach they use typically results in some blurriness. The second approach, the *off-axis, semi-spherical combiner,* includes devices like the Meta 2, which look like a futuristic pair of goggles from *Star Wars*.[4] The $1,495 device delivers relies on a single OLED flat panel built into the lenses to create an impressive 2560 × 1440 pixel display.

Another experimental method, *wavelength combiners*, revolves around a technology called *waveguide grating,* also known as a *waveguide hologram,* to produce optics. The technique taps a method called total internal reflection (TIR) to progressively extract collimated, or parallel, images within a wavelength pipe. The wavelength pipe is made from a thin sheet of glass or plastic that allows light to pass through. The diffraction that occurs in this process produces the high-quality holographic image—along with a potentially greater field of vision. For now, however, the technical challenges related to manufacturing these glasses in volume are formidable. They remain outside the realm of a commercial product. VRDs also fit into the "future" category, although Intel has announced its intention to sell smart glasses that use retinal projection.[5]

The Vision of AR

AR technology isn't just about presenting interesting, useful, or fun images. These devices increasingly tie into other smartphone or AR glass features, including cameras, microphones, and speech recognition. It's critical for glasses to coordinate various functions—in much the same way a smartphone handles the task. This requires an operating system (OS) and application programming interfaces (APIs) to connect software programs, apps, and tools. Finally, these devices require processing power along with memory, storage, and a battery. One of the persistent and ongoing problems with all portable tools—from smartphones to smart watches—is the ability to use a device for an extended period of time.

For now, no clear standard for AR glasses or HMDs exists. Researchers in labs and at technology firms continue to experiment with a variety of approaches and form factors. They're also exploring new ways to input data. This might include virtual cursive or character-based handwriting technology or gestures that use fingers, hands, head or eye movements, or an entire body. A system might also rely on a specialized stylus or entirely different types of keyboards. For example, a company named MyScript has developed an AR management technology called Interactive Ink, which supports augmented reality and other digital technologies.[6]

Technical obstacles aside, the use of AR on phones and glasses will likely soar in the years ahead. As the price of microchips and electronic components continues to drop and software and systems become more sophisticated, the technology will filter into apps used for entertainment, marketing, training, and technical support. AR will deliver information on the road, at the beach, in the classroom, and on the factory floor. A 2017 survey conducted by Evans Data Corporation found that 74 percent of mobile developers are either incorporating or evaluating the use of AR in apps.[7]

Investment and financial services firm Goldman Sachs predicts that the market for augmented and virtual reality will reach $80 billion by 2025.[8] This would put AR technologies on par with today's desktop PC market. Heather Bellini, business unit leader for Telecommunications, Media and Technology at Goldman Sachs, believes that augmented reality has the "potential to transform how we interact with almost every industry today, and we think it will be equally transformative both from a consumer and an enterprise perspective."[9]

How Virtual Reality Works

The enormous appeal of virtual reality isn't difficult to grasp. The technology transports a person into an immersive and seemingly real digital world. Designing, engineering, and assembling the technology required to produce a convincing VR experience, however, is no simple task. It's one thing to stream images to a head-mounted display or create a crude sense of feel using haptic gloves that vibrate; it's entirely another to coordinate and orchestrate computer-generated processes into an experience that convinces the mind the scene is real. This requires an understanding of how a person moves, reacts, and thinks in both the physical and VR worlds. As a result, there's a need to think of VR as a framework that extends beyond the human body.

HMDs are at the core of virtual reality. They provide the visuals and other sensory cues that allow a person to feel immersed. Unlike AR glasses or goggles, they must close off the outside world. Today, HMDs typically use liquid crystal display (LCD) technology or liquid crystal on silicon (LCoS) to generate images. In the past, HMDs relied on cathode ray tube (CRT) technology. Future displays will likely use OLED technology, which delivers clearer and brighter images. The transition from CRT to LCD and OLED has allowed engineers to build HMDs that are less bulky and more comfortable. Smaller and lighter units are also important because they allow a user to move around more freely.

An HMD requires several core components to produce the visual experience. Lenses built into the display focus a user's eyes on objects in the VR space and achieve the correct depth of field. In fact, a person's eyes don't focus on the lenses; they focus into the virtual distance. The display itself sits just beyond the lenses. It produces two slightly different images that when calibrated between two eyes produce depth and dimensionality. This stereoscopic effect contributes to the brain believing it's navigating within a 3D world. Aiding in this process are earbuds, headphones, or built-in speakers that generate spatially accurate sound. This makes the bark of a dog or the sound of a car engine seem realistic as it moves closer or further away, or across the horizontal plane.

Other technical requirements also define a VR experience. A minimum acceptable refresh rate for virtual reality is generally pegged at 60 frames per second (fps). Anything less than this refresh rate causes video stuttering that often leads to motion sickness. While high-end gaming consoles run at 60 fps, televisions operate at about 30 fps. The Oculus Go, released in early 2018, has redefined consumer VR. It runs at 72 fps and at a resolution of 2560 × 1440. This generates more realistic, smoother, and more seamless graphics. The Go also uses a technique called *foveated rendering*, which detects where a person is looking and renders these areas in full detail while rendering other areas in less detail. This allows developers to use computing resources more efficiently.[10]

The field of view for HMDs typically ranges from 90 to 110 degrees, though a few systems deliver an FOV as high as 150 or above. Like AR eyeglasses and goggles, the FOV for a system determines how authentic scenes appear to a person. Too low an FOV and a user experiences a "snorkel mask" effect that hinders the overall experience. A wider view creates a greater sense of realism and immersion. An FOV of 180 or above generally extends to the edges of human peripheral vision. Currently, the challenge for engineers is that higher FOVs result in reduced graphics resolution and sharpness.

Engineers are attempting to tackle this challenge. As the GPU chips that power these displays become more powerful and better adapted to virtual reality, FOVs will increase and, in conjunction with foveated rendering, the visual quality of VR systems will continue to improve. Newer chipsets from the likes of Nvidia and Qualcomm are advancing VR further. These chips are far more adept at handling motion, pixel shading, and other technical and practical elements that can make or break a VR experience.

Movements Matter

Tracking and navigation are also critical in VR. The Oculus Rift, for example, offers a video-game style controller called Touch that a person wears over the wrists. This provides the mechanism required to move around inside the virtual world. Newer designs, such as the Oculus Go, offer a more compact single controller with a touch surface as well as dedicated buttons. Within the VR space, the Go controller depicts a laser pointer that allows a user to navigate the interface. The HTC Vive, by contrast, offers a VR controller that not only allows a user to navigate through the VR world but also brings real-world objects into the virtual world. It also offers wrist strap controllers and special rackets for sports.

Of course, a VR system must also detect how a person moves and responds, ideally within 30 milliseconds but generally at a maximum latency of 50 milliseconds. This may include hand, body, eye, and head movements. The latter is particularly critical because the display must adjust as a person turns or tilts his or her head—including looking up, down, or backward. Some systems, such as Sony's PlayStation VR (PSVR), use a series of LED lights to track movement with external cameras. The Oculus Rift relies on a tripod system equipped with IR LEDs to track a user's movements. Software translates all this data into the scene a person experiences inside the VR environment. Other systems tap sensors—usually some combination of gyroscopes, accelerometers, and magnetometers—to track and measure movements.

The most advanced HMDs incorporate eye tracking, which delivers feedback about what a person is looking at and how he or she is responding. Foveated rendering uses a sensor in the HMD to follow eye movements. The system prioritizes what a person is specifically looking at and taps digital image processing to produce higher-resolution graphics. This also allows developers to dial down nonessential graphics and other events without a user noticing it. Using this technique, engineers and developers deliver maximum performance from the GPUs and CPUs that produce the VR graphics and audio.

Until recently, HMDs required cables to power the device and receive signals from a smartphone or personal computer. VR systems, however, are rapidly evolving into self-contained devices. Cables and cords are being replaced by on-board processing, memory, and storage, along with rechargeable batteries that last upward of 30 hours. Newer devices, such as the Oculus Go and Lenovo Daydream, operate entirely independent of other computing devices. The Go also has built in audio and the platform eliminates the need for external tracking devices. Its built-in sensors track motion and capture the position and movements of a person's head.

Researchers have also explored the idea of direct brain interfaces to control VR spaces.[11] One startup, Neurable, is working to perfect an HMD that uses electrodes and a technique called electroencephalography (EEG) to capture brain activity. Analytics software deciphers the signals and transforms them into actions.[12] In the future, this noninvasive brain–computer interface (BCI) could revolutionize the way people use extended reality. It might also allow the blind to achieve at least some sense of sight.[13]

Haptics Get Touchy-Feely

Haptic gloves are also advancing rapidly. Researchers are redefining touch technology and the user experience. The problem with many of today's VR gloves is that they provide only basic haptic feedback that's designed to mimic pressure, texture, and vibration. A slight buzz, rumble, tingle, or tug on the fingers, however, isn't like the corresponding real-world event. "These systems don't deliver a realistic analogue of what your brain expects to feel," explains HaptX founder and CEO Jake Rubin. "Smooth objects don't feel smooth. Bumpy objects don't feel bumpy. An orange doesn't feel like an orange and a handful of dirt doesn't feel like dirt. Every time an experience doesn't match the expectations of the brain the virtual-reality experience is diminished to some degree."

The HaptX system takes aim at these challenges. The gloves rely on microfluidics within a smart textile to deliver a higher density and greater displacement of feedback points. The system uses air channels with micropneumatic actuators built into a thin, flexible sheet of material to deform skin in a way that mimics touching an object. In other words, a person wearing HaptX Gloves in the virtual environment feels a virtual object as if it's a physical object. Software that manages the valves and sensors in the HaptX system translates data about pressures, textures, and feelings into a mathematical model that simulates how the human musculoskeletal system works. "The goal is to build a system that actually matches how your brain expects to feel the things you interact with in a virtual world," Rubin says.

Others are also exploring haptics technology. A Microsoft Research Team has developed a multifunctional haptic device called the CLAW that allows a user to grab virtual objects and touch virtual surfaces and receive tactile feedback as if they were real.[14] The device adapts haptic rendering by sensing differences in the user's grasp and the situational context of the virtual scene. "As a user holds a thumb against the tip of the finger, the device simulates a grasping operation: the closing of the fingers around a virtual object is met with a resistive force, generating a sense that the object lies between the index finger and the thumb," a Microsoft blog noted. The company is also experimenting with a haptic wheel and other devices that could be used in VR and AR apps.

Oculus has experimented with haptic gloves that include internal "tendons" that tense and relax. The system simulates a person's sense of touch within a VR environment. By creating the feeling of resistance—much like stretching a rubber band—a person gains a more accurate sense of touch using the haptic gloves. According to a US patent filed by Oculus: "The haptic feedback facilitates an illusion that a user is interacting with a real object when in fact the object is a virtual object. The mechanism resists movement by one or more portions of a user's body."[15] Another firm, Kaya Tech, has developed a full-body VR tracking and haptics system called the HoloSuit. The most advanced version of the product includes 36 total sensors, nine haptic feedback devices, and six firing buttons.[16]

Cracking the Code on VR

The software frameworks that support virtual reality are also advancing rapidly. SDKs are expanding virtual horizons and virtual-reality operating systems are emerging. For example, Google debuted the virtual-reality OS Daydream in 2016.[17] It allows Android smartphone users to use a wide range of VR apps from content providers such as YouTube and Netflix—as well as Google's own offerings, including Google Earth. Meanwhile, Microsoft has developed Holographic OS for VR headsets, and others are entering the VR OS space.

The number of virtual-reality apps is also growing. The Oculus platform offers thousands of app titles—from car racing and mountain climbing apps to space missions, historical events, and movies. It also offers travel apps, such as Flight over Ancient Rome and Google Earth VR with Street View. This allows a user to fly over the Manhattan Skyline, across the Andes Mountains of Ecuador, along the Nile River, and through the Grand Canyon—all with the photorealism and 3D sensation of actually being there. While a computer may generate stunning graphics, the virtual-reality space actually makes it seem real.

The Holy Grail of virtual reality is a concept called 6 DOF (six degrees of freedom tracking). It revolves around the idea that a person in a virtual-reality environment is able to move freely. This includes forward and backward, up and down, and sideways. Of course, walking blindly around a physical room—compete with walls, furniture, and other hazards—isn't a good idea for someone wearing an HMD. That's why, for virtual reality to gain even greater realism, there's a need for omnidirectional treadmills. As the name implies, the mechanical device allows a person to move in any direction at any moment. When the VR system delivers contextual cueing, walking or running on the treadmill can seem entirely natural.

Omnidirectional treadmills can be used to simulate a wide range of terrains, from hilly to flat, and they operate with or without a harness. The resulting whole-body interaction can transform the way a person perceives various stimuli and events. So far, researchers for the US Army have explored the idea, and a handful of companies has attempted to develop a treadmill that could be used commercially. One system, called Infinadeck, lets a user step in any direction and change directions in real-time. Television network CBS has reportedly explored the idea of using the omnidirectional treadmill for green-screen sets, where an actor could move around on camera in a virtual framework.[18]

Another omnidirectional system is called Virtusphere.[19] Individuals enter a fully immersive environment inside a 10-foot hollow sphere, which is placed on a special platform with a "ball" that rotates freely in all directions. When someone takes a step or changes direction the system adapts in real time. The sphere, wheel platform, and rolling bar support unfettered motion. A person in the Virtusphere wears a head-mounted display that integrates the motion through software. So far, the device has been used for training military personnel, police, and workers at nuclear power plants and other high-risk facilities that require highly specialized skills. It also has been used by museums, by architects, and for home gaming.

Another movement technology is a virtual holodeck. It allows people to move around freely in a virtual space by generating a 3D simulation of the physical space, with added elements. For example, The VOID has introduced a VR theme attraction that recreates a Star Wars film, *Secrets of the Empire,* or the movie *Ghostbusters*. The virtual space, dubbed "hyper-reality," allows participants to move around as they would in the physical world.[20] It features sights, sounds, feeling, and smell. The VOID is available at several sites, including Downtown Disney in Anaheim, California, the Venetian and Palazzo Hotels in Las Vegas, Nevada, and at Hub Zero, City Walk in Dubai, UAE. In Hong Kong, Korea, and elsewhere, other arcades are taking shape.[21] They have introduced holodecks, omnidirectional treadmills, and haptics systems for multiplayer gaming and other scenarios.

Researchers are also exploring ways to improve the CAVE. One commonly used approach involves "redirected walking," which is deployed in a CAVE space that's 10 square feet or smaller.[22] It tricks a visitor into thinking that he or she is walking along a straight line while the person is actually traveling along a curved path.[23] The VR system rotates and adapts imagery slowly and imperceptibly so that a user feels slightly off balance and adjusts his or her gait accordingly. This technology makes it possible to use the CAVE space far more efficiently.

Mixed Results

As the definition of reality is reframed and reinvented to match the technologies of the digital age, an array of experiences becomes possible. Mixed reality, sometimes referred to as hybrid reality, combines virtual and physical realities to create yet another experience in the reality-virtuality spectrum. Within this computer-generated space, actual objects coexist with virtual objects and things. At the most basic level, anything that falls between a complete virtual environment and a completely physical environment falls into the category of mixed reality, though augmented and virtual realities use mixed reality in different ways.[24] MR can be useful in a wide range of applications, including simulations, training environments, interactive product content management, and manufacturing environments.

For example, a VR environment that drops a person into a virtual conference space could also include a live person making a presentation or actual booths that are virtualized for the environment and fed into the environment via a live video stream. The conference space could also include a virtual environment with augmented overlays. This might allow a shopper to step into a virtual showroom and view different cars or appliances but also tap on them to view features and specifications. At the other end of the spectrum, a person might use AR glasses to view a scene—say, a sporting event—but then watch a virtual-reality instant replay, where he or she is inserted into the scene.

Not surprisingly, MR introduces other opportunities and challenges. Most important, all the technologies and systems used for virtual and augmented reality must integrate seamlessly. The right images, text, video, and animations must appear at the right time and they must be contextually accurate. Yet, it's also vital that the visual, haptic, and auditory experience in a mixed-reality space is presented in an easy to navigate way. As with any computing device, interface and usability are critical elements. The challenges are magnified when live video is inserted into a virtual environment or a virtual environment is embedded into a live scenario. The underlying software and algorithms must interconnect components seamlessly.

A Physical World without Borders

Virtual-reality systems orchestrate all these tasks and technologies through three types of integrated computer processing: *input*, which controls the devices that deliver data to the computer or smartphone—including a controller, keyboard, 3D tracker, or voice recognition system; *simulation*, which transfers user behavior and actions into virtual results within the environment; and *rendering*, which produces the sights, sounds, and sensations that the user experiences in a virtual environment. The latter includes haptics and auditory generation systems.

When designers, engineers, and systems developers assemble the right mix of XR technologies and optimize the way systems work, the result is a continuous feedback loop that delivers images, sounds, and haptic cues at precisely the right instant. The resulting stream of images, graphics, video, and audio puts the person at the center of the action—or displays the right data or images on AR glasses. A user feels as though events are taking place as they would in the physical world. Ultimately, the AR, VR, or MR becomes more than a new way to view objects and things. It's a framework for an entirely new way to navigate the digital and physical worlds.

The result is a new and different type of reality that extends beyond the limitations of human perception. In some cases, it may lead to *synesthesia*, a process where the right combination of electronically generated signals tricks the mind into believing parts of the body have been touched or stimulated—or some realistic seeming action or event has taken place. To be sure, augmented, virtual, and mixed reality are more than a tangle of technologies. There's a need to understand human behavior, social interactions, and how XR affects the human body and the mind. This requires a deeper understanding of the mind–body connection—and the influence of technology on everything from thinking to proprioception.

When designers, engineers, and systems developers assemble the right mix of XR technologies and optimize the way systems work, the result is a continuous feedback loop that delivers the right images, sounds, and haptic cues at any given instant. The resulting stream of images, graphics, video, and audio puts the person at the center of the action.

4

EXTENDED REALITY GETS REAL

XR Is More Than Technology

Convincing the mind and body that a virtual experience is real requires a complex mix of digital technologies. What's more, all the various software and hardware components must work together seamlessly—and continuously. Any glitch, gap, or breakdown will produce an experience that ranges somewhere between unconvincing and completely unacceptable. For example, if graphics go astray or audio isn't synced with the visual presentation, the entire experience can deteriorate into a chaotic mess. Any latency or stutter decouples the physics of XR from the physics of the natural world. In fact, research indicates that even a delay as tiny as 50 milliseconds is noticeable in a virtual environment.[1]

Within the realm of AR, things can be just as disconcerting. Hovering a phone over a menu and watching as a translator app converts French into English or German into Mandarin Chinese can lead to a few choice words if the app displays incorrect letters or words—or produces complete gibberish. Pointing a smartphone camera at a room to view what a sofa or desk will look like in the room can devolve into sheer comedy if the item floats around the room or is out of proportion. It's not good enough to simply display the object, it has to anchor in the right place and appear reasonably close to what it would look like in the room. Size, scale, colors, and perspective are all critical.

Make no mistake, orchestrating and applying a mix of digital technologies are paramount. Yet assembling systems and apps that function correctly also requires attention to other areas, including human factors and physiology, psychology, sociology, and anthropology. A person wearing AR glasses or inside a virtual world must receive appropriate cues and signals—essentially metaphors for how to interact. Part of the challenge for app designers is how to incorporate situational awareness into a continuous feedback loop. Essentially, the system must recognize and understand what a human is doing at any given instant and the human must recognize and understand what the system is doing. If a failure occurs on either side of this process, the extended reality experience will not measure up.

A common belief is that it's essential for virtual reality and other forms of XR to produce an experience as real as the physical world, but this isn't necessarily true. In most cases, the objective is to create an environment that appears real enough to trigger desired responses in the mind and body. This includes the illusion of movement or the sensation of hearing or touching an object. It isn't necessary, or even desirable, to duplicate the physical world in every way. In much the same way a person enjoys a motion picture while knowing it isn't totally authentic, an individual can suspend disbelief enough to accept the virtual-reality or augmented-reality framework as real enough.

In order to achieve a balance point, those who design and build today's XR frameworks—particularly virtual-reality software and systems—must make tradeoffs between fidelity and immediacy. Greater pixel resolution and better graphics increase the demands placed on the system to render images quickly and accurately. If the underlying computing platform can't keep up with the processing demands—as was a common problem with early VR gaming platforms—the technology unravels, performance suffers, and the magic of XR ceases to exist.

Yet, even if engineers and developers could produce an ultrarealistic virtual world, they probably wouldn't want

A common belief is that it's essential for virtual reality and other forms of XR to produce an experience as real as the physical world, but this isn't necessarily the case. In most cases, the goal is to create an environment that appears real enough to trigger desired responses in the mind and body.

to do so. The human mind and body—sensing a virtual situation is real—could wind up in a red zone. A person inside a virtual-reality environment who becomes overwhelmed might suffer a panic attack, extreme fear, motion sickness, or, in a worst-case scenario, a heart attack or other physical problem that could lead to severe illness or death.

The need to balance these factors—and embed them in the algorithms that run the software—is at the core of AR, VR, and MR. A system must avoid taking people beyond their physical and mental limits. What's more, in many cases, the VR system must adapt and adjust dynamically, based on the user's motion and responses. In the future, extended-reality systems will likely include an array of sensors and biofeedback mechanisms, possibly built into gloves and clothing, that indicate when a person is approaching overstimulation. By monitoring heartrate, perspiration, and even brain waves, systems designers can build a more realistic yet safe environment. They can offer cutoff or failsafe triggers that allow a person to enjoy the experience with minimal risk.

Perception Is Everything

It's not surprising that we take our body movements, along with the movements of the world around us, for granted.

From the moment we're born into this world we use our legs, feet, arms, hands, and sensory organs—including our eyes, ears, and mouth—to experience and navigate the physical spaces that surround us. As we grow from babies to toddlers and from children to adults, we learn the precise movements that allow us to walk, run, toss a ball, eat food, ride a bike, and drive a car. At any given moment, the combination of these calculations in our brain—including sight, sound, smell, touch, and taste—instructs our body what to do. We're essentially performing complex trigonometry without knowing it.

Proprioception refers to how we sense an environment and move our trunk and limbs accordingly. This mental and physical framework—essentially a hardwired algorithm that is largely the same for all people—allows us to do all sorts of things: eat food without seeing our mouth or even looking at the utensil; apply the right pressure to a knife while slicing carrots; pick up a pen and sign our name while blindfolded; and drive a car while changing a radio station or reaching for an item in the back seat. This "position sense," which involves the entire nervous system, allows us to handle an array of tasks "that would be otherwise impossible. The sense is so fundamental to our functioning that we take its existence for granted," according to Sajid Surve, a doctor of osteopathy who specializes in musculoskeletal medicine.[2]

Of course, during any activity in the physical or virtual worlds, our brain continuously absorbs the stimulus that is presented to it. More than 100 billion neurons and one billion synapses in the human brain continuously monitor the environment and process sensory data—collected by the eyes, ears, nose, skin, and mouth. Joseph Jerald, cofounder and principal consultant at NextGen Interactions and a member of the adjunct faculty at Duke University, points out that we "don't perceive with our hands, ears, and hands—we perceive with our brains. Our perception of the real (or virtual) world is not reality or 'the truth'—our perceptions are an interpretation of reality. ... The way to think of the situation is that objects don't actually poke into our retinas," he points out. "We instead perceive the photons bouncing off those objects into our eyes."[3]

In fact, the intersection of psychology and physiology is a critical element in VR. Studies show that humans do not perceive the physical world accurately. Frank Steinicke, a professor of Human–Computer Interaction at the Department of Informatics at the University of Hamburg, has found that people do not necessarily follow the path they think they are following in the physical world.[4] This becomes painfully apparent when people get lost and wander through forests and deserts—but hardly cover any ground.[5] In fact, with a blindfold or in a virtual environment they believe they are walking in a straight path when they are actually curving and meandering. Without feedback, people drift and cannot retrace their path. They might also believe they are walking a great distance when they are not, and they may become confused by their surroundings, even when things look familiar.

App designers must take these types of misperceptions into account when they create virtual worlds—and even augmented systems. In some cases, these spaces require virtual warping and other techniques to create the appearance of reality. This also means that with the right design and the right equipment—such as an omnidirectional treadmill, a holodeck, or a device such as the Virtusphere—a virtual environment can take up a lot less space than the virtual representation of space it encompasses. The illusion is similar to a movie set with backdrops that appear to stretch the scene into the distance when the physical boundary actually stops only a short distance away.

Of course, there are many other elements that designers and app builders must get right—for example, the way the sun shines and casts shadows, the way trees grow, or the way the horizon looks are keys to creating realism. They might be altered, but only to create a fantastical imaginary world. Yet there's a risk: if the look of a virtual space interferes with the basic physics and realities of real-world perception the space may produce disorientation, motion sickness, and cognitive distress. So, while designers can manipulate the human mind in some ways, they must also be respectful of where human limits exist. It may be possible to warp some angles and distort certain things. but taking too many liberties is a sure route to failure.

A Focus on Feelings

Ensuring that a virtual space operates correctly in the physical world is a daunting task. A great deal of complexity surrounds the way our bodies sense things—and ourselves. That's why a person who suffers a stroke or other impairment often finds previously simple tasks extraordinarily difficult. This individual must use new neural connections to relearn the task. In order to achieve convincing results in the AR and VR spaces, it's necessary for computers to process mountains of information and transform it into an interface and actions that seem natural. This requires software code—built into complex models—that allows the body to interact with the computer in real time using HMDs, haptics systems, and other technologies that create a virtual setting. Think of this as a digital form of proprioception.

Yet, despite enormous advances in VR, today's systems remain fairly primitive. Most virtual-reality designers focus on creating compelling visuals, while other sensory input such as audio and touch remain an afterthought—or wind up completely ignored. The absence of a complete array of sensory signals, however, and the resulting stimuli to the brain, essentially the framework that creates human proprioception, will likely lead to a mismatch between sensory input and output. This gap typically creates an unsatisfying, if not completely unpleasant, virtual experience.

Numerous perceptual psychology components apply to VR, Duke University's Jerald points out.[6] These encompass an array of factors, including objective vs subjective reality; distal and proximal stimuli; sensation vs perception; the subconscious and conscious mind; visceral, behavioral, reflective, and emotional processes; neurolinguistic programming; space, time, and motion; and attention patterns along with adapting methods. In a practical sense, all of these factors intersect with augmented- and virtual-reality technologies through buttons, eye movements and directional gaze, gestures, speech, and other body motions, including the ability to move around inside a virtual space.

While the complexity of all this isn't lost on anyone, the intersection of these various issues and components makes it more difficult to produce a compelling virtual space. An outstanding perceptual model isn't enough. Any environment must change in order to accommodate a nearly infinite stream of actions, behaviors, and events that occurs within a virtual application. What's more, people sometimes act and react in unpredictable ways. Translating all of this into an algorithm is vital. The task is further complicated by existing beliefs, memories, and experiences, which vary from one person to another and across cultures.

Finally, it's critical to build powerful graphics processing and display capabilities into an HMD. The software must make sense of all the motion and activity—for both the computer-generated virtual world and the person using it in the physical world—and generate a convincing virtual space. What made the Oculus Rift a breakthrough, for instance, was the use of OLED panels for each eye—operating at a refresh rate of 90 hz with ultralow persistence. This meant that any given image appears for two milliseconds per frame. In practical terms, this translates into images that display minimal blur with extremely low latency. The end result is a more convincing and immersive environment.

Although augmented reality doesn't present the same magnitude of technical or practical challenges, it isn't exempt from the laws of physics and human behavior. There's still a need to create a realistic experience that appeals to the senses—and many of the same psychological and physical components apply. What's more, depending on the display technology, the specific application and what it is attempting to do, there's a need to factor in proprioception and how people perceive the augmented space.

For example, visual clutter or a distracting interface may render the experience chaotic, confusing or dangerous.

Interaction Matters

Understanding how people act and react in an augmented or virtual space is critical. There's a need to discern interaction patterns at a detailed level. This data not only helps create a better experience, it offers valuable information about how to create an interface and maximize usability. A virtual-reality system may allow people to select objects and things through physical controls using buttons and other input systems; hand, arm, and full body gestures; voice and speech controls; and virtual selection systems, such as touching virtual objects inside the virtual space. This, in turn, may require tools, widgets, and virtual menus. Regardless of the specific approach, the computer must sense what the person is doing through sensors and feedback systems.

Interface challenges extend beyond user controls. It's also necessary to deliver viewpoint controls. This may include activities like walking, running, swimming, driving, and flying. Navigating and steering through these activities may require multitouch systems and automated tools that use sensors and software to detect what a user is doing at any given moment—or what he or she is intending to do. Consequently, designers must consider everything from the viewpoint of the person using an app to the relationship between objects and things—and place the user at the appropriate spot in the activity. In some cases, the user may control the environment while in other cases the software may aid or manage things in response to what the user is doing—or prompt the participant to change his or her behavior.

The practical result is a need for different types of input. What's more, it may be necessary to flip between input methods. For example, at the start of a virtual activity a person might use a physical button to choose an action. Later, he or she might use a virtual menu. At another point, the user might rely on a gesture or haptic feedback to simulate picking up or tossing an object. Along the way, there might also be a need for voice commands. For instance, someone might place an item in a shopping basket by touching it in the virtual space but later decide to put it back. At that point it's probably easier to say "remove" or "delete" along with the name of the item. Of course, the computer and software must react to this continuous stream of inputs and commands.

As virtual environments evolve, the tasks becoming increasingly complex—but also increasingly fascinating. One key question is how to address the challenges of locomotion. It might be possible to take a step or two in any direction, but beyond that there's a risk of bumping into physical objects, including furniture, walls, and other people. Coordinating walking or other movements within the virtual environment can involve both physical and virtual tricks. Omnidirectional treadmills and holodecks are increasingly valuable tools. But designers might also rely on teleportation techniques, such as a bird or aircraft, that transport a user from one place to another inside the virtual world. They might also include other visual and sensory tricks to create an illusion—and feeling—of movement when there is no actual movement.

Another challenge is to create the illusion that an event is taking place—say, flying an airplane or traveling in a spaceship—but avoid the sensory overload of actual g-forces (the pressures created by acceleration and rapid motions). The need for a muted experience is obvious. This technique helps reduce sensory conflicts that lead to dizziness or motion sickness. Engineers have a name for this concept: *rest frames*. It refers to a pause or stationery moment that helps the user gain a sense of spatial perception. For instance, a race-car driver or spaceship pilot would normally feel the vehicle rattling, vibrating, and pitching as it moves at high speed. Inside the virtual world, the car or spacecraft doesn't move. Instead, the outside environment flashes by at the speed a race car or spacecraft would travel while the participant receives other sensory stimulation, such as vibrations and sound.

Rest frames rely on a basic but important principle: they combine a dynamic field of view with a static field

Within a virtual world, it may be necessary to create the illusion an event is taking place but avoid the sensory overload of actual g-forces and motions. The need for a muted experience is obvious. This may help reduce sensory conflicts that lead to dizziness or motion sickness inside a virtual world.

of view. In a practical sense, this means that parts of the display are in motion while other parts are not. By making objects fade in certain frames, it's possible to trick a user's mind. Not surprisingly, there's a delicate line between too much and too little stimulation in a VR space. Users, and particularly those playing games, desire an adrenalin-provoking experience. What's more, teleporting, while valuable for certain situations, isn't always desirable, because it reduces or eliminates part of the virtual world. Dynamic rest frames create a sense of balance because they deliver the excitement and immersion of the activity without creating mental and physiological overload.

An added benefit of dynamic rest frames is that they can adjust to the user's movements and inputs dynamically. Because of this, they can be used to direct focus in different directions. This is especially valuable for cuing users about which way to look or how to navigate a space. Yet, this isn't the only visual and sensory trick VR developers use. They also rely on a simpler method called *shrinking field of view*, which involves focusing more narrowly on what the participant sees. The technique is widely used in games and it appears in old silent movies. In some cases, developers may use both techniques, though dynamic rest frames deliver a far greater feeling of stability.

A Touch of Reality

Another key to creating a realistic virtual experience is haptics. Producing a virtual replica of an actual tactile event is no small task. The human body experiences the sense of feel in complex ways. Nerve endings in the skin feed information about pressure, heat, and other sensations to the brain, which instantaneously decides what the feeling is and what it means. This helps humans understand how to act and react within a physical environment, such as when a coffee cup is too hot to hold, or a silk fabric caresses the skin. To be sure, these sensations can lead to pleasure or pain. This feedback loop is referred to as the *somatic sensory system*.

A haptics system must take numerous factors into account. The human hand consists of 27 bones connected by joints, tendons, muscles, and skin. This physiological structure delivers what is known as 27 degrees of freedom (DOF). This includes four in each finger, three for extension and flexion, and one for abduction and adduction; the thumb has five DOF, and there are six DOF for the rotation and translation of the wrist.[7] Consequently, humans can press, grasp, pinch, squeeze, and stroke objects. When a person touches or holds an item, two basic types of response occur: *tactile*, the basic sense of contact with the object, and *kinesthetic*, which involves the sense of motion along with pressure. As the nervous system transmits signals to the brain, a person constantly adjusts to handle the task at hand.

Mapping this level of functionality to a digital environment requires sensors, artificial muscles and tendons, and sophisticated software. Joysticks and other basic computing input devices cannot deliver more than two or three DOF. Other haptic devices that have emerged from research labs have improved the DOF factor to six and occasionally seven. By contrast, HaptX technology delivers tracking accuracy to about a third of a millimeter at each finger with six DOF per digit, 36 DOF total in the hand. This means that a person using the system can engage in a wide array of tasks that involve detailed motions at virtually any axis.

It's not necessary to duplicate the exact physiology of the hand. "The system tricks the brain into thinking you're touching something real," HaptX CEO Jake Rubin says. The technology is much more precise than systems using motorized components. In the future, the technology could be built into full body suits and exoskeletons that create sensation beyond the fingers and hands. Used with high-fidelity motion tracking and other systems, such as an omnidirectional treadmill or specialized chamber, a person could virtually experience activities from the physical world. This might include parachuting out of an airplane, deep sea diving, or rock climbing up a sheer cliff.

A Sense of Scent and Taste

The quest to create realistic virtual reality also leads to scent and flavor. This might include an ability to smell or taste a dish at a restaurant before setting foot in the eatery or training a HAZMAT worker to identify potentially dangerous fumes. It might also allow a travel agency to create a virtual lavender field in Provence or a patisserie in Paris so that potential travelers can gain a sense of what the actual experience might be like. Of course, by combining scent and taste with sight and touch it's possible to reach a major goal of virtual reality: a completely immersive and convincing experience.

Technologies that produce virtual scent are advancing. These include specialized oils, chemicals, and pellets that are typically activated by pressure or electrical signals and released into a scent chamber.[8] Stanford University's Virtual Human Interaction Lab, for example, has concocted virtual doughnuts that look and smell real.[9] A company called FeelReal has built a sensory mask that produces scents while also mimicking other sensory stimuli, including heat, mist from water, vibration, and the feeling of wind. The company touts an ability to generate the "smell of gunpowder" while playing a VR game, or an ability to "smell flowers" while roaming through a 3D virtual garden. The wireless sensory mask, which fits over head-mounted displays from Oculus Rift, Sony, and Samsung, generates other scents from a base of chemicals. This includes a jungle scent, burning rubber, fire, and the ocean.[10]

Researchers are also looking to incorporate taste into virtual environments. For example, at the National University of Singapore, a group of researchers has developed a digital lollipop that can emulate different tastes.[11] The goal, according to research fellow Nimesha Ranasinghe,[12] is to build a system that allows people to taste a recipe they see on a cooking show on TV or on the internet. The technology uses semiconductors to manage alternating current and deliver slight changes in temperature to a user. This approach fools the taste receptors in the mouth—or more specifically the brain—into believing that an electrical sensation is a taste. The same research group is also exploring ways to transmit tastes over the internet using electrodes.

Out of the Lab and into the Virtual World

The real-world challenges of putting all the pieces together and creating immersive lifelike applications are enormous. Virtual objects must be able to move but also rotate, scale, and change positions. What's more, a user must be able to select objects and do things with them. This requires sophisticated controllers as well as software that can manage the data stream and render the environment correctly.

While a realistic experience is important for a computer game, it's nothing less than critical for training a police officer how to diffuse bombs or helping a surgeon understand exactly how perform the cuts required for an operation.

A trio of researchers at Walt Disney Imagineering noted in a 2015 academic paper that the potential of extended reality is inhibited by the sheer number of factors that must be addressed. This includes "the lack of suitable menu and system controls, inability to perform precise manipulations, lack of numeric input, challenges with ergonomics, and difficulties with maintaining user focus and preserving immersion."[13] The goal of system designers, they noted, "is to develop interaction techniques that support the richness and complexity required to build complex 3D models, yet minimize expenditure of user energy and maximize user comfort."

Part of the problem is that commercial off-the-shelf and widely used desktop modeling tools for VR, such as Maya and SketchUp, do not provide the wealth of spatial information required to create robust virtual environments. Some tools use a 2D interface to create a 3D space. Further complicating matters, while there is plenty of evidence that 3D interaction offers a better way to build 3D virtual environments, most software in this category lacks the robustness of 2D toolkits. It's a bit like trying to draw a detailed portrait of a person in a computer using a mouse rather than a digital pen. Lacking the right tool, it's impossible to achieve the level of detail that makes the portrait a good piece of art.

There is another fundamental challenge: legacy input devices—including game controllers, wands, and other devices with buttons, triggers, and a joystick—were never designed to operate in a 3D virtual world. This inhibits—and often limits—the way input and design take place. The end result is a VR environment that is somewhat awkward. "This can limit the expressiveness of user interaction due to the simplistic nature of these types of inputs while simultaneously complicating the interface due to the large number of inputs and the need for the user to remember complex functional mappings," the Disney researchers wrote.

The Disney team has focused its attention on developing a hybrid controller that collocates a touch display, offers a casing with physical buttons, and delivers the ideal of six degrees of freedom. They have also focused on developing more robust software that allows a user to select tools and widgets within the virtual space. Their stated goal is to produce a navigation system—complete with touch screens and floating virtual menus—that enables interaction across a spectrum of screens and displays, from desktop PCs to head-mounted displays and CAVE environments.

Others, such as the startup Nuerable, are working to advance brain–computer interfaces to the point where the computer uses direct neurofeedback from the brain to eliminate controls altogether. This requires a greater understanding of brain signals—and an ability to reduce the noise from the brain and in electronic systems—in order to produce a system that can operate in real time. “We believe that science has reached the early stages of a neurotechnology revolution that will eventually bring BCIs to everyday life. Neurotechnology holds tremendous potential to assist and augment human cognitive functions across a wide variety of tasks,” the company says on its website.[14]

Still other researchers and other commercial firms are exploring completely different ways to take virtual reality to a more realistic level. Google has experimented with light-field photography—a system that captures all the rays of light from a scene rather than only what passes directly through a camera lens—to produce more realistic virtual graphics.[15] The system uses a 16-camera circular rig with GoPro cameras positioned in a semicircle to capture more dimensional data about the image than a single camera can achieve. The goal is to produce “beautiful scenes in stereoscopic 360 VR video,” Google noted. These sights, whether they depict the Incan ruins of Machu Pichu or the International Space Station, can be viewed on high-end HMDs as well as inexpensive Google Cardboard displays.

Virtually There

In an ideal virtual world, a person picks up an AR or VR application and uses it without a manual or any guidance. The interface is intuitive while the computer graphics are completely believable. This goal is the same whether the representation is a realistic place or a fantasy world that could never exist in the physical world. For app developers and digital artists, this may require new plants, animals, and creatures; introducing different types of avatars; and producing entirely new virtual settings and ways to move around that defy the physics of the real world—flying or swimming under the ocean, for example—while adhering to the physics of movement. Digital artists must also think about how to focus attention and redirect the brain and senses through light, patterns, and sensory cues.

Virtual environments ultimately hinge on two critical factors: *depth of information* and *breadth of information*. This framework, introduced by computer scientist and communications expert Jonathan Steuer, revolves around the idea of maximizing immersion and interactivity.[16] Depth of information refers to the richness of the environment as defined by display graphics, resolution, audio quality, and overall integration of technologies. The breadth of the environment involves the spectrum of senses engaged by the overall VR technology platform.

In practical terms, this means that an AR or VR application that operates accurately 90 percent of the time isn't good enough. A character that doesn't respond correctly even one out of five times in a game, or one that periodically disappears from view for a few seconds, is enough to undermine the entire experience. Likewise, an augmented overlay that serves up a hodgepodge of data—or the wrong graphics and information—may confuse an engineer or technician and lead to an error. Getting to the 100 percent level in performance, usability, and plausibility, however, extends beyond current technology and knowledge. For now, the goal is to create a space or app that's simply plausible. If a person can suspend his or her sense of belief, the application works.

Of course, advances in digital technology guarantee that future systems will be far more sophisticated—and components will be more tightly and seamlessly integrated. Pixel by pixel, frame by frame, and app by app, augmented reality, virtual reality, and mixed reality are changing the world. In fact, these virtual technologies are already reshaping almost every industry and sector—from manufacturing to entertainment and from engineering to medicine. Says Marc Carrel-Billiard, global senior managing director of Accenture Labs: "The world is moving from a flat screen to immersive 3D."

5

VIRTUAL TECHNOLOGIES CHANGE EVERYTHING

Extended Reality Will Redefine Work and Life

Although virtual technologies have been around in one form or another for a few decades, the hype has mostly exceeded the reality. During the 1990s and beyond, XR experienced one false start after another. Vendors have consistently overpromised and underdelivered for products and features. Graphics were mostly subpar, performance was jittery, and, in the end, the user experience ranked somewhere between mediocre and awful. It usually didn't take long for the novelty to wear off and a headset or glasses to wind up collecting dust on a shelf or tucked away in a desk drawer.

Virtual technology is finally taking shape, however. As systems and software have advanced and the hardware has shrunk, AR, VR, and MR have begun to filter into daily life. Rachel Ann Sibley, a futurist, consultant, and part of the faculty at Singularity University, believes that almost every corner of the world and every industry will be disrupted by virtual technologies in the years ahead. “There are a lot of reasons why we have not yet seen mass consumer adoption,” she stated at a Trace3 Evolve conference in 2018. Yet, “every leading technology company is vying for leadership in this space.”[1]

What makes extended reality so disruptive and so potentially valuable? Sibley believes that it is the technology platform of the future. AR, VR, and MR put data, images, and objects into shapes and forms that are natural and understandable. In fact, she refers to augmented and virtual reality as the “connective tissue” of the future. Extended reality bridges the psychological chasm between machines and humans. “Every physical object can be abstracted, demonetized, democratized, digitized,” she explained. So far, most of these digital objects—cameras, calendars, address books, voice recorders, and specialized apps—have resided in our smartphones.

But 2D is not 3D. A flat screen is not immersive, no matter how beautiful or brilliant the display. Simply put, a smartphone can't create a complete sensory experience. Extended reality takes computing and digital consumption in an entirely different direction. AR, VR, and MR fundamentally change the way we learn, shop, build, interact, and entertain ourselves. These technologies produce new ways of thinking by rewiring sensory processing. They make things possible that once seemed unfathomable: touching a rare fossil, swimming with sea turtles, soaking in the grandeur of Niagara Falls, and experiencing the magnificence of the Taj Mahal, Eiffel Tower, or Lincoln Memorial.

Let's take a brief tour of how extended reality will likely play out in the world and how it will shape and reshape several industries and disciplines.

Education

The internet has already changed education in remarkable ways. It hasn't replaced physical or virtual books, which remain a powerful tool for critical thinking. But it has enabled distance learning, spawned new types of online interactions, and introduced entirely different ways to view and explore data and information. At some point, however, human interaction is essential. Although online spaces are ideal for allowing users to sift through materials at any time and from almost any place—while progressing at their own pace—the downside is the lack of intuitive interaction that takes place in a classroom. Online learning doesn't deliver a particularly dynamic experience.

AR and VR take direct aim at this challenge. Suddenly, it's possible to experience the familiarity of a real-world

It's clear that virtual technologies will fundamentally change the way we learn, shop, build, interact, and entertain ourselves. They will open up new vistas of thinking—and sensory processing. They will also make things possible that once seemed unfathomable: touching a rare fossil, swimming with sea turtles, exploring a spaceship, and experiencing a famous edifice such as the Taj Mahal, Eiffel Tower, or Lincoln Memorial.

school without traveling to a campus and setting foot in an actual classroom. Using avatars, virtual representations of physical objects and places, and graphical interfaces and models, participants can step past 2D tools like messaging, chat, video, and whiteboards and experience a lecture, discussion, or virtual field trip in a lifelike way. They can view things—from molecules and mathematical formulas to interstellar objects—in shapes and forms that make sense. Ultimately, the technology allows people to interact in a more natural and seamless way.

Academic studies support the notion that AR, VR, and MR are effective tools for enabling learning. Extended reality also makes learning more appealing and enjoyable. For example, research conducted at the University of Saskatchewan in Canada found that the use of VR improved learning accuracy by about 20 percent for medical students studying spatial relationships.[2] When the research team tested participants five to nine days after undergoing instruction, those who learned through virtual reality scored higher than those who used textbooks. "A week later it seemed like I was able to go back into my mind and bring back the experience," said one participant in the study.[3] Likewise, a 2018 study conducted by researchers at the University of Maryland showed an 8.8 percent improvement in overall recall accuracy.[4]

Extended reality introduces opportunities to expand learning from kindergarten through college—and beyond. Students might visit a zoo, step onto the surface of the moon, view the signing of the Declaration of Independence, stroll through graphical representations of molecules in their chemistry lab, or watch a volcano erupt in Iceland. They might join virtual lectures and engage in virtual discussions. Already, a handful of schools are experimenting with AR, VR, and MR technologies. For instance, the Washington Leadership Academy, an open-enrollment public charter high school in Washington, DC, is reinventing learning through digital technology, including extended reality.[5] The initiative was partially funded by Laurene Powell Jobs, the wife of former Apple CEO Steve Jobs.

Training and Career Development

An enormous challenge for businesses of all shapes and sizes is training employees. In the United States alone, companies spend upward of $70 billion annually.[6] It's no secret that skills change rapidly. Critical knowledge a few years ago is now out-of-date. What's more, as digital technologies become more pervasive in all industries and lines of work, there's a growing need to learn new skills. Although online training has revolutionized the way many workers update their skills and knowledge, it has never delivered the realism of assembling a jet engine, mastering a sales technique, or installing a flow control valve at a nuclear power plant.

Augmented reality presents opportunities to enhance training by putting manuals, technical specs, and other data on glasses or lenses as a person learns or performs work. Virtual reality reduces, if not eliminates, the risks associated with many types of training. It's possible to be prepared to walk into a structure that's on fire or deal with terrorists holding a hostage before stepping into a real-world situation. Moreover, individuals gain important skills before working with valuable equipment or supplies. The fact that VR removes the need for physical classrooms and transporting people to a specific spot at a specific time is no less appealing.

One company that understands the value of XR is British aircraft engine manufacturer Rolls-Royce Holdings. It has turned to virtual reality for far better results in manufacturing.[7] Rolls-Royce uses the technology to teach workers how to assemble critical components in jet engines used on aircraft. A worker dons a VR headset and then steps through the entire process of placing specific components and parts in a gearbox. If the technician makes a mistake, the system issues a warning and forces the technician to start over. Once a worker learns the right sequence and placement for parts, it's time to move on to the actual assembly station and work with actual engines.

There's also the ability to hold conferences and symposiums in virtual worlds and mixed-reality spaces. Accenture is developing a teleportation system that allows participants to view a virtual representation of a speaker, view avatars for other participants as they ask questions, and virtually walk through spaces and view booths and exhibits that are part of the physical world. "As bandwidth has improved, video quality has gotten better, and digital technologies have connected to one another, we have advanced online meetings from simply connecting to other people to more advanced telepresence and now virtual and augmented realities," explains Emmanuel Viale, a managing director at Accenture Labs.

In fact, Viale and a team of researchers are designing the conference room of the future from a lab in the south of France. They have combined scanned and mapped 3D models of physical things—walls, spaces, and objects—with immersive virtual spaces so that 10, 20, or more people can congregate to view presentations, watch videos, listen to lectures, ask questions, and interact. Everything—video, audio, avatars, and movements in the space—are synced in real time. There is also an element of augmented reality within the virtual-reality environment. It's possible to view names, titles, and information about avatars that represent participants and also view data about objects visible in the space. This includes everything from videos to people and things.

Accenture researchers are also exploring other ideas. One of them is virtual autographs. In the VR environment, participants would capture data about another person, with his or her permission, and the system would automatically recognize and reproduce the avatar or representation in the future. Another idea is creating a framework to support people in the physical world interacting with people in the virtual world, without both groups being together in the same space at the same time. In order to make all of this possible, some participants would wear headsets and venture into an immersive virtual environment, while others would wear smart glasses or lenses—or use a smartphone—to interact. The technology blends all elements and groups into the same digital space.

Other mixed-reality environments are emerging as well. Microsoft's SharePoint Spaces allows participants to share documents, images, videos, and 3D graphics within a 360-degree shared virtual space that's visible on a web browser. The application, which doesn't require a special headset or glasses, lets participants interact in a more lifelike way on their PCs. The platform might help new employees explore an office or a campus, or a business team view data in more visual and understandable ways. Microsoft offers the environment within its popular Office 365 platform.

Sports

In 1998, the National Football League (NFL) introduced augmented reality to telecasts. The initial use of the technology was to help viewers see first down lines that are not visible on a physical field. The "First and Ten" system, which features an augmented yellow line projected onto the field, has become a standard feature for televised NFL games. In recent years, television networks have added virtual scoreboards along with graphics that show fans the anatomy of plays, replays, stats, and other information in formats that make a telecast more compelling and enjoyable.

Other sports leagues are also transforming telecasts through AR features. The National Basketball Association (NBA) introduced Virtual 3 in 2016. It highlights the three-point line when a player takes a shot outside the arc. Sports network ESPN has also experimented with AR for major league baseball (MLB) games. This includes a "K-Zone that shows balls and strikes. The objective, in all these cases, is for fans to have a better idea of what's going on in a game, particularly when it's something that's difficult to spot with the human eye. Not surprisingly, many of these AR features borrow techniques that first appeared in video games.

Virtual reality is also scoring points in the sports world. VR creates opportunities to insert fans into the action—and view plays from various angles and perspectives. It introduces ways to view graphics and information that aren't possible on a TV screen or at a ballgame. Consequently, professional sports leagues, including MLB, the NBA, and the NFL, have begun broadcasting select games and highlights in virtual reality. In addition, at the 2016 Summer Olympics in Rio de Janeiro, Brazil, NBC and the BBC streamed some events in virtual reality. At the 2018 FIFA World Cup, all 64 games were available to viewers in VR via an app from the television network Telemundo.

VR is still a work in progress in the sports world, however. The biggest problem, for now, is that the video resolution in virtual-reality environments doesn't match the most advanced high-definition television sets that display images at 4K. Many viewers have described the VR experience as incredible but the graphics as underwhelming or disappointing. Televising games in VR also requires substantial human and technical resources. For instance, a virtual-reality NBA game broadcast in 2017 required a crew of 30, a TV production truck, banks of cameras, and three announcers.[8]

Analysts have speculated that in the future VR could be used for college recruiting—allowing a recruit to visit schools, view facilities, and even check out a weight room or arena, without actually setting foot at the school. It could also become part of an athlete's training regimen.

Golden State Warriors guard Stephen Curry, a league MVP and NBA champion, announced in June 2018 that his personal trainer was working on a three-minute pregame drill that uses VR. "We always look for new things, just to keep him stimulated and to keep pushing his workouts forward," his personal trainer, Brandon Payne, stated.[9]

Engineering, Manufacturing, and Construction

Architects, engineers, and designers were among the early adopters of augmented and virtual reality. They now use XR and 3D modeling tools for visualizing spaces as well as prototyping things as diverse as buildings and automobiles. This allows them to see what something will actually look like before it's built—and spot flaws that could lead to cost overruns, design errors, or safety hazards. AR and mixed-reality tools also allow a person to step through a virtual space—say an office or museum—and view technical specs and information about electrical systems, plumbing, and other elements. Wearing a pair of AR glasses during the construction phase, data and graphics are superimposed over an actual physical view of the room or object.

The technology boosts productivity and cut costs. It allows organizations to operate in a more agile and flexible way. "There are an enormous number of industrial uses for extended reality," points out Mary Hamilton, managing director of Accenture Labs in San Francisco. "These technologies introduce capabilities that simply didn't exist in the past. They fundamentally change and improve processes."

Aerospace giant Boeing has experimented with Google Glass Enterprise Edition to streamline assembly processes in its factories. Typically, aircraft require assembly of tens of thousands of individual components, including wires that come in myriad shapes and sizes. In recent years, engineers and assemblers have used PDF files to view assembly instructions. However, using a keyboard while engaged in the assembly process is tedious and slow. Navigating a keyboard also increases the risk of error. Instead, the AR system delivers specific contextual information about how to find specific wires, cut them, and install them.

Although Boeing has experimented with heads-up displays in the past—the earliest pilot projects at the company extend back to 1995—Google Glass helped the initiative get off the ground. Technicians authenticate themselves by simply putting on the glasses and scanning a QR code. They immediately view detailed wiring instructions and other crucial data in a corner of the eyewear. The app that drives the system, Skylight, incorporates voice and touch gestures. This makes it possible to glide through steps seamlessly and ensure that the right instructions and assembly phase are visible at any given moment. Glass and Skylight helped workers decrease the assembly time by 25 percent while significantly reducing error rates, according to a Boeing executive who oversaw the pilot project.[10]

Ford Motor Co is also using VR to build automobiles. It has created an immersive vehicle laboratory. In 2013, the auto manufacturer examined 135,000 items on 193 vehicle prototypes without building a physical model.[11] In the future, AR and VR will be used across industries for assembly instructions, maintenance, technical support, quality assurance, and much more. One device, DAQRI Smart Glasses,[12] essentially brings a control panel to a person's eyes. This means a worker can roam through a factory and handle tasks and functions as the need arises. It's possible to video chat, view critical specs, review 3D models, and manage digital controls hands free.

Entertainment and Gaming

Extended reality is already reshaping the way consumers view videos and movies and participate in games. In 2016, the augmented-reality game Pokémon Go captured the hearts and minds of millions of people around the globe. Suddenly, participants combed backyards, street corners, parks, and other public places in search of virtual creatures with names like Bulbasaur, Charmander, and Squirtle. When all was said and done, the AR treasure hunt—which resembled nothing before it—resulted in upward of 800 million downloads worldwide.[13] More important, it set the stage for a tsunami of AR games that followed along with more widespread acceptance of augmented reality.

Immersive virtual-reality gaming is also advancing. In Catan VR, for example, the popular board game becomes a rich interactive 3D experience. Objects, characters, and avatars emerge from the virtual board, scurry about, and engage in a variety of actions and activities. In Ark Park, gamers enter a spectacular futuristic fantasy world where they ride and hunt dinosaurs. Star Trek: Bridge Crew assembles participants from all over the world. People take on different tasks—navigating, operating the engines, managing the weapons—as the USS *Aegis* explores uncharted sectors of space in the pursuit of finding Vulcans a new home. Along the way, there are dangerous objects and attacks from Klingons.

Gaming and entertainment are also taking on new and different forms. In 2016, for instance, a movie theater in Amsterdam became the world's first permanent VR movie cinema,[14] in which theatergoers plunked down €12.50, plopped themselves into a rotating chair, and strapped on Samsung VR headsets along with a pair of headphones. They watched a specially made 35-minute film and, by rotating their seat or turning their head, they could view the virtual reality action in 360 degrees. Although the theater shuttered its doors in 2018 the concept lives on at pop up VR cinemas across Europe. In Japan, cinemas also show films in virtual-reality formats.[15] Mainstream studios such as Paramount and theater formats such as IMAX have experimented with VR as well.[16]

The capabilities of virtual reality are clearly expanding. For example, VR "theme park" The VOID thrusts visitors into a Star Wars battle on a molten planet and recover "intelligence" vital to the rebellion's survival. Participants, wearing helmets, gloves, and special vests, see, feel, and hear everything around them as if they were actually in a physical space. Specialized hardware and software—including haptic gloves, motion tracking systems, and built-in special effects—make it possible to move around in the holodeck. Without goggles, the rooms are essentially bare walls and stages. With the headset, they become completely immersive spaces. A door becomes an elevator; a hallway becomes a bridge to a spaceship. There are controls, enemy storm troopers, and monsters at every turn.

Consumers are also exploring virtual reality in arcades and from the comfort of their sofas. With an HMD such as the Oculus Go, PlayStation VR, or HTC Vive, it's possible to ride an extreme roller coaster, battle an army of homicidal robots, race cars through a rocky desert landscape, fly spaceships, embark on a treasure hunt for hidden gems, and even play popular games. What's more, visual programming tools such as Microsoft's AltSpaceVR, help users design and build their own virtual worlds—and share them with others.[17] This essentially adapts the concept of the popular online space Second Life to the 3D virtual world.

In the coming years, as graphics and displays improve, haptic gloves advance, body suits and exoskeletons emerge, and more advanced motion tracking takes shape, ultrarealistic immersive gaming will become the new normal. People will also attend virtual plays and music festivals that, in some cases, take on novel forms.

Travel

The travel industry is also likely to change significantly as a result of VR and AR. By the end of 2018, worldwide online travel sales had reached about $700 billion.[18] And while websites and apps provide a wealth of information about destinations and getaways—they allow consumers to book flights, cruises, resorts, hotels, and tours—it can be difficult to know what a place is really like before physically setting foot there. There's also the fact that a lot of people prefer to travel vicariously, or, because of physical limitations, cannot travel to faraway places.

In a virtual space, it's possible to explore the glaciers of Greenland, the Buddhist temples of Bhutan, the emerald cliffs of Kauai, or the diverse wildlife of the Galapagos Islands. It's also possible to stroll through resorts and hotels and have a much better sense of what swimming pools, beaches, and rooms look like before reserving a room. A virtual-reality brochure or journey also allows a person to explore a region or destination in a more personalized and intuitive way. A click of a pointer on a map sends the visitor to that virtual place. Once there, selecting menu items, say the birds of the Galapagos or the chants of Bhutanese monks, makes the experience come alive.

Virtual reality has already unleashed change in the travel industry. For example, in 2015, Marriott Hotels introduced an in-room VR travel experience called VRoom Service that displays VR Postcards—essentially immersive travel experiences—using Samsung Gear virtual reality. This made it possible to experience an array of exotic places.[19] In 2016, German airline Lufthansa introduced VR kiosks at Berlin's Schönefeld Airport. Passengers passing through the terminal and awaiting flights used an Oculus headset to embark on a virtual tour of Miami or the Great Barrier Reef in Australia.[20] Others, such as National Geographic and Smithsonian, have introduced immersive travel experiences that span the globe.

The Smithsonian and the Great Courses, for example, have introduced an interactive journey to Venice, Italy, that includes a gondola ride and visits to the Grand Canal, the Piazza San Marco, Marco Polo's house, and the Basilica di Santa Maria della Salute.[21] A virtual adventurer sits in a small boat with a professor of Italian history, Kenneth R. Bartlett, who delivers a private tour for the virtual journey. The boat pitches and sways while other gondolas and boats pass by. A turn of the head reveals a 360-degree view of the scene—including the spectacular structures surrounding the canal and the gondolier at the back of the boat.

Others are stretching the technology further—and even creating virtual museums that make art, culture, and history more accessible. In 2018, a firm called Timescope introduced the idea of self-service virtual-reality kiosks that virtually transport users to places all over the world via a telescope-like system equipped with a 360-degree rotation mechanism, directional speakers, and 4K screen resolution. A visitor controls the system with a multilingual touchscreen that transports a person to an immersive environment such as the Monument des Fraternisations in France. There, it's possible to view the trenches of World War I during a cease-fire between Allied and German forces.

AR also changes the way people travel and interact with others. Smartphone apps such as Google Translate already deliver instant translations for signs, menus, brochures, and other printed items. Apps such as Looksee use AR to display places and things, as well as distances for major sights, over live views on a smartphone screen. It includes cities such as Barcelona, Paris, Los Angeles, and Orlando. Meanwhile, at Gatwick Airport in the UK, an award-winning app projects a path—displayed as a series of green arrows—to guide a passenger to the correct gate;[22] a user simply holds his or her phone up and the path is projected onto the screen with the live physical space behind it.

Journalism and News

Few industries are likely to undergo as big a change as the news media in the coming years. For better or for worse, there's a growing emphasis on news as a "spectacle" and form of entertainment than a source of information about the world. In March 2018, CNN announced that it was introducing a virtual-reality news app for the Oculus platform.[23] The app allows viewers to experience events in an immersive way—and at far higher resolution than ever before offered in the VR world. A person can view what it's like to be on an aircraft carrier when a jet takes off, in a helicopter during a rescue or in the middle of a battle, on a camel in the middle of the Sahara Desert, or at the center of a festival or sports event. The CNNVR app introduced 360-degree 4K video digital content with a news ticker at the bottom of the screen. It also included integration with social media, such as Twitter.

Other news outlets, including *Huffington Post* and the *New York Times*, have also begun to produce AR, VR, and MR content. Some of these experiences put a viewer at a sports event or rock concert; others involve stepping into refugee camps and crime scenes. While societal standards have broadened about what's acceptable to show in a video, the issue takes on new meaning when the scene is ultrarealistic and immersive. Some, like Robert Hernandez, an associate professor at the University of Southern California's School for Communication and Journalism, have stated that VR could invoke bad memories and even unleash trauma for some people. The role of journalists and the news media may require reexamination. "As a journalist I have to ask what's my job? Is my job to hurt you? It's to inform you, and sometimes it's a little in the middle," he stated.[24]

Marketing, Retail, Shopping, and Real Estate

One of the first uses of AR in marketing occurred in 2008. A firm in Germany created a printed magazine ad of a model BMW Mini that, when held up to a camera on a computer, appeared on screen in 3D.[25] A virtual model connected to markers on the physical ad made it possible for a user to control the paper version of the car and move it around to view different 3D angles on screen. This real-time digital interaction demonstrated the potential of augmented reality. Other brands, from National Geographic to Disney, have since embraced the technology to depict things as diverse as environmental problems and cartoon characters interacting with people on the street.[26]

Other companies are expanding the concept of XR in other ways. IKEA's augmented-reality app that allows users to preview a piece of furniture has garnered press attention and consumer interest. Sherwin-William's paint app that lets a buyer view the color of paint in a virtual representation of their bedroom or office has proven popular. Others are now joining the XR party. For instance, cosmetic company Sephora has turned to AR technology to allow customers to see what various products look like—from lipstick to eyeliner—on their digital face. A user simply snaps a photo in the app or submits it to a website, selects a product, and instantly views the results.[27] Sephora Virtual Artist also offers tips that coincide with the products.

In the future, an HMD will allow individuals to venture online and view items more authentically, walk through a shopping space in a more realistic way, and, through the use of haptics and feedback technology, feel the texture of a silk blouse or a sofa. In fact, virtual reality could deliver online shopping experiences that resemble a physical store.

Instead of viewing clothing or power lawnmowers on a 2D screen, a person might travel down a virtual aisle and gaze at items—while pulling up specs and information by reaching out and touching them or directing a virtual laser pointer at them. AR and VR will also likely revolutionize real estate by allowing people to walk through property from across town or from the other side of the world. One firm, Matterport, already offers hardware and software that creates 3D virtual home tours.[28]

The concept of virtual shopping hasn't escaped retailing giant Walmart, which filed a patent in August 2018 for a virtual-reality showroom that lets users view images of shelves and products and select them virtually. Other retailers are also dialing into the concept.[29] In 2017, home improvement giant Lowe's introduced a virtual-reality shopping experience called Holoroom.[30] Customers could view VR tutorials of home improvement projects and gain hands-on experience. During a trial run of the initiative, Lowe's reported that customers demonstrated 36 percent better recall of how to complete a do-it-yourself project. The company also gained valuable feedback about where customers typically become confused and frustrated.

Auto manufacturers such as Audi and BMW are also steering toward VR. An app from Audi, for example, allows a prospective buyer to view and explore the interior of a vehicle in 3D—using a VR headset at home. In 2016, Audi introduced an app that allows customers to view different colors and options using an Oculus Rift or HTC Vive HMD at a dealer. Now it's expanding its virtual-reality offerings to users at home. Using an app developed by a firm named ZeroLight, potential buyers can enter a virtual cabin and see what the dashboard, seats, and space look like in any of the manufacturer's 52 models. Shoppers can also view a vehicle's engine and, using headphones, hear what the car sounds like.

Law Enforcement and the Courtroom

Augmented reality could dramatically change policing. Although law enforcement officials in China already use facial recognition glasses to identify suspected criminals, the technology—in conjunction with the Internet of Things (IoT), has other applications, including license plate recognition, identifying chemicals in a bomb or other explosive device, displaying vitals when first responders encounter injured individuals, enhanced night vision, and capturing video with GPS stamps to record and authenticate events. Virtual reality is equally attractive for police. In Morristown New Jersey, for instance, officers train in virtual-reality simulators. This helps them learn when and where to shoot—and when to refrain.[31] VR could also aid in recruiting and allow the public to experience what it's like to be a police officer on a beat.

There's also the issue of AR and VR used in the courtroom. These technologies could lead to new types of evidence and allow juries to see and experience events in entirely different ways. Instead of viewing scratchy and blurry CCTV images, jurors could be transported to a virtual crime scene. Caroline Sturdy Colls, an associate professor in forensic archaeology and genocide investigation at Staffordshire University in the UK, has stated: "Traditional means of documenting, sketching and photographing crime scenes can be laborious and they don't provide data suitable for presentation in court to non-experts. A number of novel, digital, non-invasive methods used in archaeology, computing and games design present the opportunity to increase search efficiency and accuracy and provide more effective means of presenting evidence in court."[32]

In 2016, Colls received a £140,000 research grant from a European Union commission to study the use of virtual-reality head-mounted displays for the criminal system. Using an HMD, a juror could walk around a crime scene and view it from different angles and in different ways. In 2018, she received another grant to study how officers can better analyze buried and concealed evidence at crime scenes. She and other researchers are collaborating with UK police forces to take policing and courtroom trials into the digital age.

Medicine and Psychology

Virtual reality is also introducing new ways to treat physical and psychological conditions. At Cedars-Sinai Hospital in Los Angeles, psychiatrists have experimented with VR to treat opioid addiction. Results show the approach is promising.[33] At the University of Maryland, doctors use AR to view ultrasounds while examining a patient.[34] This makes it possible for a technician to avoid looking away from a patient while conducting the exam. It also puts essential information in view of the technician at all times. Another use for the technology comes from a company called AccuVein, which has developed a device that scans veins and projects an AR visualization onto the skin area a technician scans.[35] Using the handheld unit, it's possible to view veins inside the body. The process has led to 45 percent fewer escalations, the company claims.

Another firm, Surgical Theater, has introduced a VR platform that allows patients to view their anatomy and doctors to view tumors, blood vessels, and other structures in the brain or another part of the body.[36] The system displays a patient's unique issue and helps doctors plan the operation. Physicians at George Washington University Hospital in Washington, DC, already use the technology, which it calls Precision Virtual Reality.[37] Apart from reviewing information and planning the procedure with much greater insight, they can use the personalized 3D

Virtual reality is also changing medicine. At Cedars-Sinai Hospital, psychiatrists have experimented with VR to treat opioid addiction. At the University of Maryland, doctors use AR to view ultrasounds while examining a patient. This makes it possible for a technician to avoid looking away from a patient while conducting the exam.

virtual images to demonstrate to patients what will take place.

Meanwhile, a San Diego, California, organization called the Virtual Reality Medical Center has turned to VR to aid people with phobias like fear of flying, public speaking, agoraphobia, and claustrophobia. VRMC uses 3D virtual-reality exposure therapy in combination with biofeedback and cognitive behavioral therapy to treat phobias, anxiety, stress, and chronic pain.[38]

Pharmaceutical and biotech companies are also tapping virtual technologies. Jonas Boström, a drug designer in the Department of Medicinal Chemistry at AstraZeneca in Sweden, has developed a molecular visualization tool called Molecular Rift, which runs on Oculus goggles. It offers an environment where a person can use gestures to interact with molecules and examine how they behave in different environments and situations. Boström believes the tool is "the next generation of molecular visualization."[39] Other pharma and biotech firms rely on the CAVE and other VR tools to engage in advanced drug discovery. This includes virtual mock-up studies, design reviews, safety studies, ergonomic studies, failure mode effects analysis, training, machine redesign assessments, computer-aided engineering, and air flow visualization.

Virtual reality will also aid in detecting and diagnosing dementia and other problems, including PTSD.[40] At

Rush University Medical Center in Chicago researchers have developed modules to help medical and pharmacy students recognize signs and symptoms of dementia.[41] At the University of Southern California, Skip Rizzo, who has worked on the BRAVEMIND project, is exploring ways to treat veterans and others suffering from post-traumatic stress disorder. He has built simulated environments where people "relive" traumatic events in a virtual space that resembles a computer game. The technique builds on a widely accepted treatment approach called "prolonged exposure" therapy. The initial versions of the VR program, called Virtual Iraq and Virtual Afghanistan, were adapted from the 2004 Xbox video game *Full Spectrum Warrior*. He later developed the more advanced virtual-space BRAVEMIND as a tool for helping PTSD patients confront and process difficult emotional memories.

These VR environments feature a variety of combat situations, conditions, and wound levels. At any time, a psychologist can trigger a reenactment of the original traumatic event. This approach clearly shows promise. A study conducted by the Office of Naval Research in the United States found that 16 of 20 participants displayed significant reductions in PTSD symptoms as a result of the VR treatment.[42] One soldier indicated that reliving his traumatic experiences in a virtual environment substituted the need to think about the trauma when he was at home with his family. Researchers have also used virtual environments to aid victims of sexual trauma and other types of abuse.

Rizzo believes that virtual reality will eventually emerge as a mainstream tool for addressing both physiological and psychological issues. Today, "physical rehabilitation for a stroke, traumatic brain injury or spinal cord problem requires a person or robotic device to deliver resistance," he explains. "The next frontier is introducing VR simulations that can gauge a person's physical capacity and apply pressure and resistance exactly how it is needed to rehab the person." Instead of actually touching or picking up physical objects, for example, a system will send signals and auditory cues that result in a believable experience. "You reach for a balloon and it pops when you squeeze it. You hear the sound, you feel the air and you experience the vibration." Further out, he says, systems may include exoskeletons that simulate walking and allow patients to engage in activities that aid in rehab or allow a person with a disability to experience the sensation of walking, running, or swimming.

The Military

It's shouldn't be the least bit surprising that the military and organizations like the US Defense Advanced Research Projects Agency (DARPA) are among the biggest adopters of AR and VR tools. The quest for superiority on the battlefield has led to massive research and development efforts in the XR space. Much of the effort has revolved around training and simulations. Teaching pilots how to fly or soldiers how to react to dangerous situations is a lot less expensive—and far safer—in a virtual world. A simulator that costs $800,000, such as the US Army's Stryker,[43] is more practical—and effective—than dropping tens or hundreds of millions of dollars into actual war games and training exercises.

Virtual-reality simulators also collect valuable data about how soldiers and others act and react in different situations. Using artificial intelligence and analytics tools, it's possible to spot patterns and trends that lead to more effective combat methods. Yet, the use of XR extends beyond simulators. For example, the US Army has developed a head-mounted display that projects relevant data, information, and graphics into a soldier's view—while providing a view of the physical battlefield.[44]

In the future, these systems could also include biodetection capabilities and they could integrate with sensors that might be worn in clothing or carried on a person. Across the Atlantic Ocean, the British Army has introduced a virtual-reality recruiting tool that allows potential recruits to experience training exercises. It witnessed a 66 percent increase in signups after offering the immersive experience at various sites across the UK.[45]

XR Remaps the Physical World

Extended reality is filtering into other parts of life as well. A pastor in Reading, Pennsylvania, is working to create a virtual church, complete with avatars, music, and sermons.[46] New York's Museum of Modern Art has introduced augmented reality to a Jackson Pollock exhibit. With a smartphone in hand and an app called MoMAR Gallery, a visitor sees the art radically remade along with illustrations that completely revamp the space.[47] Even rock bands and fashion designers are dabbling in the space. They're introducing new ways to experience their designs, music, and other creations. While VR might prove valuable for viewing traditional paintings and sculpture, it will likely usher in completely new forms of art—including remarkably different installation art—in the years ahead.

Of course, as these various forms of extended reality take shape—and reshape the way people go about their daily lives—a multitude of questions, concerns, and potential problems arise. The psychological and sociological changes extended reality unleashes intersect with ethics, morality, legality, and behavior. While some of these issues are merely intriguing, others veer into the category of somewhat to extraordinarily disturbing.

6

MORALITY, ETHICS, LAW, AND SOCIAL CONSEQUENCES

A Change in Values

Every new technology unleashes changes in human thinking and behavior. For example, the ability to capture images onto film allowed people to view photographs of family members—and view exotic places and things that they wouldn't otherwise see. It was a transformative moment in history. More than a century and a half later, the digital camera allowed people to snap photos without worrying about film or the cost of processing it. This fueled social media and led to dramatic changes in the way people interact. Suddenly, with a smartphone in hand, it was possible to share images—including videos—instantly.

This evolution in photography led to changes that were both predictable and unintended. The latter includes outcomes and consequences that aren't always desirable—including things as diverse and problematic as revenge porn and the viral spread of fake images and memes. As AR, VR, and MR advance and grow in popularity, a similar scenario is unfolding. The impact of these virtual experiences extends far beyond the pixels, sounds, and physical sensations that an artificially induced reality produces. Virtual technologies reshape personal thinking and societal attitudes on a massive scale.

All of this raises questions and concerns that extend into and across myriad fields: psychology, physiology, sociology, anthropology, philosophy, law, ethics, and even religion. *How will people adapt to countless hours of immersion? Will some people become cyber-addicted and opt to remain in an immersed state as their preferred state? What would this do to the dividing line between the physical and virtual worlds? How does virtual technology affect hacking and virtual criminality? What is the impact of virtual porn and sexuality, which could take on entirely new dimensions in the virtual world? And how will the legal system address all sorts of new issues and problems?*

Designing Responsibly

Fear, uncertainty, and doubt have swirled around almost every new technology. In the early 1800s, the Luddites opposed the use of any and all weaving machinery. Their protests and uprisings lead to the destruction of weaving machines and textile mills across England. In 1877, a *New York Times* editorial railed against the telephone as an invasion of privacy. And in 1943, IBM chairman Thomas J. Watson predicted that the world market for computers would top out at about five.[1] Three years later, 20th Century Fox executive Darryl Zanuck predicted that "television won't be able to hold onto any market it captures after the first six months. People will soon get tired of staring at a plywood box every night."[2]

Although it's easy to laugh off past predictions gone wrong, it's also wise to recognize that emerging technologies represent the unknown—and all the unpredictability that comes with it. Yet, somewhere between hype and fear lies the real world of how various inventions influence and change people and societies. Printing presses, telephones, cameras, televisions, computers, and every other technology along the way altered how people go about their daily lives and engage in activities. Augmented and virtual realities will follow the same trajectory. Designers, manufacturers, software developers, and others will make decisions that affect the way people use and experience these systems. Along the way, society will have to decide what's desirable, what's acceptable, and what's off limits.

Clearly, virtual reality raises plenty of concerns. For starters, if an environment is too realistic, a person within the immersive world might experience dizziness, nausea, disorientation, panic, or even a medical problem such as a stroke or heart attack. Although it's reasonable to think a person having a bad VR experience would simply yank off the headset and reunite with the physical world, things aren't so simple. The mental anguish associated with being shot at or dropped from an airplane without a parachute could prove so terrifying that a person succumbs to an immediate panic attack and freezes. There's also the possibility of long-term psychological impacts, including desensitization.

In fact, visitors to virtual spaces say that the feelings and sensations often seem entirely real. A 2016 *Wall Street Journal* article highlighted software worker Erin Bell, who, while standing on a carpet with a VR headset strapped on, refused to step off a plank suspended over a deep, rusted pit—even though she knew she was standing in a lab at Stanford University and a researcher was initiating the request. "I knew I was in a virtual environment but I was still afraid," she said. Others have reported strong feelings of anxiety, fear, and dismay as scenes and experiences unfolded within a virtual world. These range from driving a race car at high speeds to stepping into virtual combat. What's more, research shows that when people enter a virtual reality space they suspend belief in the physical world. Simply put, their minds tell them the experience is completely real.[3]

Virtual reality raises plenty of concerns. For starters, if an environment is too realistic, a person within the immersive world might experience disorientation, panic, even a medical problem such as a stroke or heart attack.

Although manufacturers engineer VR systems to minimize nausea, dizziness, and other negative sensations—and most recommend that young children do not use these systems—there are also growing concerns about eyestrain, headaches, repetitive stress injuries, and other physical effects. Oculus, for instance, offers the following recommendation: "Ease into the use of the headset to allow your body to adjust; use for only a few minutes at a time at first, and only increase the amount of time using the headset gradually as you grow accustomed to virtual reality." Oculus doesn't stop there: "Take at least a 10 to 15 minute break every 30 minutes, even if you don't think you need it. Each person is different, so take more frequent and longer breaks if you feel discomfort."[4] It also warns about driving, riding a bike, or using machinery after a VR session.

Mind over VR

USC's Skip Rizzo is among the world's foremost experts in virtual reality and the human mind. His research has focused on developing VR technology to aid and treat PTSD, spinal cord injuries, and other conditions, which we discussed previously. The technology is a two-sided coin, however. "There are many questions that remain unanswered," he says. "Virtual technology can be used to alter emotions and treat people with PTSD and other conditions. But if we accept that VR can be used to evoke emotions for a positive aim, then we also have to accept that it could also trigger negative emotions that could have a long-lasting effect."

In fact, the long-term effects of virtual reality are largely unknown. It's reasonable to think that—as with today's gaming devices—extended immersion could lead to antisocial behavior and other types of psychological problems. Research conducted in the virtual reality space indicates that the technology changes the way people think and behave. The mind, tricked into thinking the artificial events are real, acts as though they are. This, of course, leads thinking and behavior in new directions. While the outcomes aren't necessarily negative, they also aren't necessarily positive. Indeed, it's possible for a user to land anywhere along a broad continuum of outcomes.

Stanford University's Jeremy Bailenson believes that the impact of virtual reality is discernable and significant. He has examined VR and human thinking for nearly two decades. The lab builds and studies systems that allow people to meet in virtual spaces and explore new and different types of interactions. "Qualitatively, it's got a different impact on how we perceive information because we're using our bodies," he stated.[5] "It is simply not the same as going from regular resolution to HD TV, or from black and white to color videos. Because you're using your body and it's completely immersive, I do think it's a big jump in the history of media. But I don't think it's going to change who people are."

In Bailenson's book *Experience on Demand*, he points out that a person in a virtual world can step into another person's shoes to feel what it is like to do a job—or experience what it's like to be homeless. Bailenson is particularly interested in how VR can help people develop greater empathy and understanding for others. He has conducted extensive research on ageism, racism, and aiding the disabled. He writes: "If a teenager has a negative stereotype of the elderly, then merely asking the teenager to imagine an elderly perspective might simply serve to reinforce those stereotypes. The teen might create images of elderly who are slow, frugal, and who tell boring stories."[6]

Adding VR to the mix could prove transformative. While certain types of training can desensitize people to other people's feelings—and this may be a good thing for a soldier or firefighter who must witness atrocities and calamities—the technology can also play a positive role. "Formalizing the role-playing into a set simulation that counteracts stereotypical representations can avoid these negative stereotypes by creating scenes that show the strengths of the elderly," Bailenson noted. In other words, participants might lack information that's crucial to gaining real-world perspective, and VR can guide them through the process. He goes onto say: "Empathy isn't a fixed quality. Our capacity for empathy can be changed by our culture and the media technologies that transmit the culture's values—for good and bad."[7]

How VR affects empathy is highly a nuanced topic, Bailenson points out. For example, in one experiment, participants were twice as likely to help those who are color-blind after spending time in a VR environment that allowed them to see the world as if they were color-blind themselves. However, VR can also produce counterintuitive results. Another study, for example, found that individuals who became "blind" in a VR world sometimes became more discriminatory and less empathetic.[8] The reason? The newly blind person experienced the trauma of sudden blindness rather than the ongoing reality of being blind. Simply put, they are focused on the obstacles related to coping with the problem rather than the problem itself.

The ramifications are important and must be considered by those designing applications as well as those using them. There are no simple or clear answers to how the mind reacts to virtual stimuli. Virtual reality "is not a magic bullet," Bailenson writes. "It doesn't succeed in its mission every time, and the 'effect size,' the statistical term for how strong an outcome is, varies."[9] Rizzo adds that "cognitive reappraisal" resulting from VR can take many directions, both positive and negative. "For now, there is a lot of research and exploration going on," he says. "Psychology as a science has been around for at least 125 years, studying how humans behave and interact in the real world. We now have to begin the process of studying how humans behave and interact in the virtual world and try to understand the implications of this and its impact on life in the real world."

Changes in thinking aren't limited to building empathy and gaining a better understanding of others, however. At the heart of VR is a basic concept: it's possible to be anyone—including a different version of ourselves. "When one sees her avatar—whether it's from the first person looking down at her digital body, or by looking in the virtual mirror—and events occur synchronously, her mind takes ownership of the virtual body. The brain has very little experience—from an evolutionary standpoint—at seeing a perfect mirror image that is not in fact a reflection of reality. Because we can digitally create any mirror image we want in VR, VR offers a unique, surreal way to allow people to take ownership of bodies that are not their own," Bailenson writes.[10]

Essentially, VR exploits the malleability and adaptability of the human brain to convince a person that the avatar or virtual body is real. A mirror image that is slightly different—or even entirely different—becomes one's self when the senses are fully engaged. This "body transfer" allows a person to become a solider, a firefighter, a police officer, an astronaut, a deep-sea diver, or even a bird soaring above the earth, and accept the make-believe role or identity as real. It also allows an individual to change genders, experience different social roles, feel what it's like to be a different race or religion, and experience life as a baby or a 90-year-old.

A similar phenomenon is the principle of "uncanny valley." The concept was introduced by robotics professor Masahiro Mori in 1970 as *Bukimi no tani genshō* (不気味の谷現象).[11] It refers to the human inclination to accept a robot or virtual being as more human the more real it looks—up to a point. With this comes greater empathy. Oddly, after the artificial representation or robot reaches a certain threshold of realness, the human response switches to revulsion. When the robot switches back to a little less human, the feeling of attraction and empathy return. One of the interesting takeaways from Mori's discovery—beyond basic design ramifications—is that it introduces the ability to psychologically manipulate humans.

Imagine a product advertisement or a political ad that shows the desirable item or virtual image of a person in a positive way and the undesirable item or person in a negative way. A human inside the virtual space may not be able to consciously discern the difference. Such an approach could also be used to manipulate thinking about news events. The 2016 US presidential election set off a heated debate about the role of “fake” or “false” news. A Buzzfeed News poll that year found that fake news headlines fool American adults about 75 percent of the time.[12] Many describe current methods of deception primitive, however. One of the most disturbing aspects is the ability to produce realistic-seeming videos of people saying and doing things they didn’t say and do. These so-called deepfake videos can depict politicians making statements they never made, insert celebrities into porn videos, and show people committing crimes they didn’t commit. Factor in virtual reality and things quickly become a lot uglier.

New Dimensions of Virtual Gaming

No space epitomizes the positives and negatives of extended reality more than gaming. On one hand, a person is able to experience places, events, and things that would otherwise reside only within the realm of imagination. This might include driving a Formula One race car through the narrow streets of Monaco or participating in the invasion of Normandy during World War II. Gaming on a 2D device like a phone, television, or computer is certainly alluring. Gaming in the immersive world of 3D virtual reality elevates the experience to an entirely different level. It raises questions about how realistic the game should be and what impact it has on a user over the short-term and the long-term.

Of course, many of these questions—and the associated issues—revolve around electronic gaming in general. There has been much written and spoken about desensitizing society to violence. Studies indicate that this isn't an abstract matter. The American Psychological Association reports that violence can have a harmful effect on people, particularly children.[13] Psychologist Craig A. Anderson, a distinguished professor and director of the Center for the Study of Violence at Iowa State University, has concluded that "the evidence strongly suggests that exposure to violent video games is a causal risk factor for increased aggressive behavior, aggressive cognition, and aggressive affect—and for decreased empathy and prosocial behavior." In fact, Anderson has found that playing violent video games increases aggressive thoughts, feelings, and actions both online and in the physical world.[14]

Yet, researchers don't unanimously agree that computer simulations are dangerous or destructive. Patrick M. Markey and Christopher J. Ferguson, in their 2017 book *Moral Combat: Why the War on Violent Video Games Is Wrong,* argue that media and politicians railing about the deleterious effects of video games, particularly after every tragedy, is completely misguided.[15] Among other things, they point out that data doesn't support these conclusions and the same arguments show up in different form with different technologies in different eras. At one point, television was viewed as a cauldron of violence and immorality. Over time, that was extended to comic books, rock music, computer and video games, and now virtual reality. As one bogeyman fades a new one takes shape.

Markey, a professor of psychology at Villanova University and Ferguson, a professor of psychology at Stetson University, offer some insight into why this scenario takes place. They wrote: "In real life, the consequences are so dire that violence—or the threat of it—causes us anxiety, and so we avoid it. But our brains seem to understand that media violence is different. Violent video games provide us with the same intensity and stimulation as walking down the dark alley, but without the anxiety associated with the possibility of actually being murdered. Our brains become excited by the graphic violence and the competition in Manhunt as if it were real, but because we also understand that it is not, we feel thrilled rather than frightened."

Grand Theft Auto is often cited as an example of all that's wrong with computer games. It's focused on stealing cars and committing an array of other crimes. Over the years, it has served up a litany of controversial elements: prostitutes; torture; full frontal male and female nudity; the power to murder other characters, including individuals in specific ethic groups; and the ability to go on dates and have consensual sex with virtual characters. Yet, the ultrarealistic graphics of the console version of the game pale in comparison to virtual reality. Suddenly, characters advance from 2D to 3D—and the person in the virtual game feels a deeper sense of realism that can be both thrilling and unnerving.

Convincing viewers that a computer game is real, or at least real enough to be fun or exciting, is a cause for celebration—but also concern. One thing that has made graphic and violent computer games more acceptable is the fact that events taking place on the screen—a murder, a bad accident, or a monster approaching the viewer—are somewhat abstract threats that typically affect others. The same can be said for movies that display an array of shocking scenes and events. While the player or viewer may become absorbed by the action, he or she doesn't feel a direct and palpable threat. The physical distance between the screen and the eyes creates a buffer zone that protects the mind of the person experiencing the scene.

But what happens in a virtual-reality game when, as Erin Bell at the Stanford University lab learned, the cyberworld seems completely real? Suddenly, things are not happening to other people, they are happening to oneself. Imagine being robbed at gunpoint and then shot. Imagine sprinting toward an enemy in war and suddenly encountering a bayonet in the abdomen, complete with a sensation of being stabbed. Imagine a car wreck that looks impressive when it's happening to someone else in a game but isn't as impressive when you're sitting in the driver's or passenger seat—especially if you have experienced a crash in the past. Imagine a monster standing over you and threatening to tear you apart limb from limb—or witnessing a friend torn apart. The fear level—and the resulting physiological impact—might be ratcheted up by an order of magnitude. At some point, the pleasurable aspects of mild fear tip over to extreme or anxiety-provoking terror.

Questions and concerns multiply and become even more complex within multiplayer games. Suddenly, participants—who connect by avatar and have no physical contact with others in the gaming environment—have the power to inflict virtual harm on others. Studies show that when a person doesn't know the person he or she is interacting with there's a much higher likelihood he or she will dole out aggressive behavior.[16] Conversely, when a person knows someone, he or she tends to display more empathy.[17] *Once a virtual-reality environment incorporates all the senses, how realistic should the scenario be? What level of pain should be allowed and how much pain should people feel? Should a new rating system be created, especially for children?* There are also psychological aspects that cannot be ignored. As USC's Rizzo points out: "For one person the environment may prove cathartic and help diffuse violence in the physical world. For another, it may incite violence in the physical world."

Studies show that presence and anxiety are closely linked, although not always in a bad way. In some instances, stress can actually lead to greater pleasure and focus.[18] Yet researchers say that it's important to pay close attention to potential negative impacts, particularly when people play games or engage in stressful virtual activities for prolonged periods. Marty Banks, a professor of optometry at the University of California, Berkeley, has found that long-term VR can cause serious eyestrain, a condition he calls "vergence-accommodation conflict."[19] Researchers at Temple University report that "cybersickness" and other maladies aren't uncommon.[20] Denny Unger, CEO and creative director at Cloudhead Games, warned that a high level of realism could lead to heart attacks and deaths. "You really could kill somebody," he said at a 2014 conference.[21]

It's critical to recognize that while people react differently to the same stimuli, in the end we're all human beings, Rizzo says. "The common denominator is that our eyes, ears and limbic systems are wired pretty much the same way. It's not only your eyes and ears, your body is a participant in virtual reality." What's more, once omnidirectional treadmills and full body exoskeletons are introduced into gaming and other VR offerings, the stakes will certainly increase. "None of these things are deal breakers but they are all factors that designers must pay close attention to when they are creating games and other virtual environments," he explains.

Virtual Porn and Sex

It has been said that sex makes the world go 'round. The Dead Kennedys, a popular rock band in the 1990s, took the concept a step further by penning the lyrics: "Kinky Sex Makes the World Go 'Round." Whatever your views on sex and kink, one fact is abundantly clear: humans, technology, and sex are inexorably linked. "The adage is that if you want to know where technology is going, follow the porn," wrote Damon Brown in the 2008 book *Porn & Pong: How Grand Theft Auto, Tomb Raider and Other Sexy Games Changed Our Culture*.[22]

In the early 1900s, French postcards typically displayed women in seminude or nude poses. These postcards were extraordinary popular, despite the fact that it was usually illegal to send them through the mail. Along the way, movies began to incorporate scenes with nudity and sex. The 1896 silent film *Le Coucher de la Mariée* showed a woman performing a striptease. It has been labeled the first documented case of pornography.[23] In the United States, the 1933 film *Ecstasy* featured a bare Hedy Lamarr. One of her films reportedly depicted a female orgasm on film for the first time.[24] By the 1960s, filmmakers were churning out scores of theatrical and pornographic movies that tested societal boundaries on nudity and sex.

Then, in the 1980s, computer gaming took off. As people plucked Atari 2600s and other gaming consoles from the shelves of electronic retailers, they not only loaded Pac-Man and Mario Brothers onto their machines, some paid $50 or more for porn games that consisted of little more than crude graphics ported to a gaming cartridge. As the decade unfolded, titles such as *Bachelor Party* and *Custer's Revenge* appeared. The latter game allowed users to rape a native American woman who was tied to a post.[25] Not surprisingly, the game—like many in that era—found itself in the crosshairs of protestors. Nevertheless, scores of new games oozing with nudity and sex followed—further extending the boundaries of porn.

Today, all of this may seem a bit passé. The internet is rife with images and videos displaying virtually every kink and fetish imaginable. Nudity and porn have gone mainstream. The next frontier, of course, is virtual reality. Not only does VR make images appear more realistic, it adds an in-your-face dose of sensory stimulation. For example, a 2015 application called Your Neighbors and You delivers a virtual-reality porn romp that lets a participant experience sex with three women. They stare into the user's eyes, they talk dirty, and they perform various acts. An April 2018 *Wired* magazine story, "Coming Attractions: The

Rise of VR Porn," quoted a man named Scott, who downloaded and used the application. "I just couldn't believe the immersion level that it provided. Even though it was a little fuzzy, everything made me realize that this is more than just watching a video in 3D. When a woman comes up close to your face, you can feel the heat coming off of her, you imagine that you feel her breath. Your brain is tricked into sensations that aren't there because of the ones that are there," he explained.[26]

Your Neighbors and You is only the entry point into an elaborate porn labyrinth brimming with fantasies and possibilities. Full body suits and exoskeletons that deliver tactile sensation within VR environments could produce stimulation and sensations in every part of the body. The porn industry has already recognized the possibilities—and opportunities. It is eager to take things to the next level. Not surprisingly, this raises important questions. *Will virtual porn become a substitute for actual sex? Will it spur virtual prostitution? Will the idea of a partner who doesn't resist, regardless of the request, be more appealing than a partner who doesn't go along with the fantasy? And what happens when the scenario involves controversial behavior, such as choking, slapping, or sexual assault? What happens when a pedophile wants to have sex with children in a virtual environment?*

There are no simple or obvious answers. Some researchers believe that VR porn could be more human-like and less objectifying than conventional porn. This might make it more appealing to broader audience, including women, who account for about one-quarter of visitors to porn sites. It might also help young men view sex in a more realistic and respectful way. But the ethical aspects of the equation hinge on society—and the legal system—to decide what's acceptable and what's legal. "At some point, somewhere, somebody will build a child molestation interaction scenario that will be reprehensible by most peoples' standards," Rizzo says. While the thought is disturbing, he points out that such apps might actually be legal. Possession of photos that depict a real child being exploited are certainly illegal; however, a virtual environment that is simply a graphic representation may be another matter.

In fact, the US Supreme Court ruled in 2002 that virtual child porn that does not include the exploitation of actual children is protected by the First Amendment.[27] The Court struck down a law from 1996 that had outlawed virtual representations of child porn. Yet, virtual reality introduces a new twist: high-fidelity graphics that look so real they are nearly indistinguishable from photographs. Rizzo says that the next challenge—if such content appears—is whether these types of experiences encourage or prevent the exploitation of children.

"It will certainly become an ethical morass as there are a lot of gray areas regarding sexuality and whether the acting out of sadomasochistic fantasies in a simulation should be viewed in the same way as they are viewed in the real world," Rizzo says. The complexities of the situation are further magnified by multiplayer or multiparticipant virtual-reality environments that will almost certainly appear in the coming years. "These are issues that an ethical society must address as VR becomes more realistic and available."

In 2013, a stage play, *The Nether*, explored an array of troubling questions that straddle the line between fantasy and reality.[28] Playwright Jennifer Haley introduced a character named Papa who created a virtual reality fantasy world where he could molest and kill young children. The play presented a number of chilling ethical questions, including whether it's possible or desirable to regulate or control personal thoughts that involve heinous acts. Perhaps the only thing that is clear at this point is that the ethical complexities of addressing pornography and the regulation of the medium will only increase as virtual reality takes shape.

Hacks, Attacks, and Privacy

On any given day, venture online and you are likely to encounter news about a security breach. Individuals, businesses, and government are under constant assault. Virtual reality will introduce new—and sometimes frightening—challenges. Today's online scams could take on new dimensions and risks if the person asking for a credit card number or bank account number appears to be a friend or family member who wants to send you a birthday gift—but is actually a crook. Without strong authentication methods, there might be no way for a user to know whether the person is who they claim to be.

It's also not unreasonable to think that a hacker could break into a system and reprogram an application to do something unexpected or terrifying. Imagine playing a game or being immersed in a virtual travel app—say you're walking down a street in San Francisco—and you suddenly encounter a mugger who isn't supposed to appear in the sequence. He walks up, pulls a gun, and demands money. If you don't pay, he fires the gun, unleashing a haptic blast of unpleasant sensations. In a 2D world, it might be shocking. In an immersive 3D virtual world, this might be enough to cause serious physical or psychological trauma.

There are also questions about virtual criminality. Cybercriminals have already found ways to launder money using game currencies and virtual currencies such as Bitcoin.[29] In fact, it's also not uncommon for individuals participating in multiplayer online games today to find out they've been hacked and that their possessions and virtual currency have been stolen. *What happens in a virtual world when a person has money and possessions stolen?* These might include vehicles, furniture, avatars, and an array of other items. Moreover, hackers and attackers might break into a system, insert or rearrange code, and unleash mental anguish, pain, or physical harm on participants. What may be annoying and frustrating in a computer game played on a 2D screen could become hurtful and destructive in a virtual world.

Still another concern involves privacy and protecting personal data. While many of the same issues apply to today's online environment, the stakes are ratcheted up further when it's possible for marketers and others to monitor eye movements, involuntary facial gestures, and other behavioral responses—and collect user data that could include medical problems and sexual preferences. Recent battles between government agencies and various web hosting firms and internet services providers haven't done anything to alleviate anxiety over privacy. *Should companies like Oculus and Sony be allowed to track users' behavior within VR spaces? Should agencies and law enforcement officials have access to records and data about a person's proclivities or behavior? What if the person participates in a virtual world where rape and child porn occur? And what about marketers and others who might abuse highly personal data?*

Legal Matters

At some point, morality crosses a line into legality. For now, there is no such boundary in the virtual world. But there are plenty of important legal questions that will play out in courtrooms over the next decade—and beyond. Eugene Volokh, a law professor at UCLA, and Mark A. Lemley, a law professor at Stanford University, have explored the topic in recent years.[30] "Virtual Reality (VR) and Augmented Reality (AR) are going to be big—not just for gaming but for work, for social life, and for evaluating and buying real-world products. Like many big technological advances, they will in some ways challenge legal doctrine," they noted.

Inside virtual worlds, a number of activities could come under scrutiny, including street crimes such as disturbing the peace, indecent exposure, and dishing out deliberately harmful visuals or other stimuli. There are also unresolved questions about tort lawsuits involving various entities in the virtual world. *What happens when there's a dispute about who owns particular avatars and likenesses? What happens when a person impersonates someone else and commits a crime?* Further complicating matters, *what happens when a crime occurs in the same virtual space but different physical jurisdictions—and the rules and laws are different in different states, provinces, or countries?* Society may need to "rethink just what constitutes a legally binding contract and what things we want governed by public rather than private rules," Volokh and Lemley write.[31]

The practical considerations are intriguing—and somewhat disturbing. For instance, *what happens when a person enters a public space with numerous people and someone is accosted or groped?* Although this may seem abstract by today's online standards, it's actually something that has already taken place. Volokh, in a 2017 story for the *Washington Post*, describes an actual woman who was virtually accosted in a multiplayer game. Another person dished out menacing gestures, including "making grabbing and pinching motions near my chest," she recounted. "Emboldened, he even shoved his hand toward my virtual crotch and began rubbing."[32]

Current laws wouldn't classify a virtual grope as a crime; there's no sexual assault or battery because the person hasn't actually been touched. In a virtual space, however, where people can impose their virtual touch on others through haptics, the act takes on an entirely different context. Likewise, *what happens if a person imposes a heinous act in the presence of others, including children?* "We base many rules on the distinction between the mental and the visceral, between things we perceive and things we experience," Volokh and Lemley noted. "VR and AR will make it harder to draw those lines, and it may push us to think hard about why we punish certain kinds of conduct and not others in the physical world." It might also lead to protections systems in virtual worlds, including repelling or blocking undesirable avatars.

Ultimately, this framework may force society—and the legal system—to "reassess and rethink the notion of what is 'real' in a world where more and more of our most significant and emotionally real experiences aren't 'real' in the classic physical understanding of that term. But they feel real, and they can have real physiological consequences," the pair wrote.[33] Volokh and Lemley cited research conducted by psychologists that indicates behavior sometimes dismissed as an "emotional" injury has physiological repercussions. These range from post-traumatic stress disorder to the damage done in an abusive relationship.[34] Further complicating matters, cultural definitions and values vary around the world.

The Human Disconnect

Although augmented and virtual reality may benefit people with physical and behavioral problems—it could prove to be a boon for treating an array of conditions—it could also lead to physical and mental health problems. Nausea, dizziness, eyestrain, and body strain aside, there's the risk of repetitive stress injuries and real-world injuries from walking into walls or falling into things with an HMD blocking the view of the physical world. The fact that VR

Although augmented and virtual reality will benefit people with physical and behavioral problems—it could prove to be a boon to treating an array of conditions—it could also lead to physical and mental health problems.

systems typically block out external sights and sounds means that it's far more difficult to remain aware of one's surroundings. What's more, warning devices built into systems don't always deliver accurate results—or alert a person fast enough.

The underlying psychological framework for participants is also a source of concern. Some of those who wear a virtual-reality headset for a period of time report that they wind up dreading the idea of returning to physical reality. Tobias van Schneider, an avid user of VR, posted a disturbing account of how his mind reacted to virtual-reality technology in a 2016 blog post. "Using a VR headset and especially playing room scale experiences is magical," he wrote. "What I quickly noticed after a couple hours of intense VR sessions is the feeling you get the hours after. ... What I'm talking about is a weird sense of sadness."[35] He complained that colors weren't as bright and experiences were far less intense.

Some have referred to this outcome as a post-VR hangover. A 2006 study conducted by a group of researchers, including Frederick Aardema, research assistant professor in the Department of Psychiatry at the University of Montreal, found that virtual reality increases dissociative experiences and lessens people's sense of presence in actual reality.[36] The researchers also found that a magnification effect takes place. "The greater the individual's pre-existing tendency for dissociation and immersion, the greater the dissociative effects of VR," they reported. Other studies have found that feelings of unreality in the physical world can be triggered by contradicting natural sensory input. This occurs because there's a mismatch between the signals the brain receives and what the inner ear and brain process.

There are other concerns, too. Addiction is already a problem in the gaming world. There are countless stories of people who can't hold jobs, stay in relationships, or manage friendships because they are hooked on gaming. There are also documented cases of people playing games for days on end—without eating or going to the bathroom—until they drop dead. For instance, in 2015, an 18-year-old man in Taiwan died after a gaming binge that lasted three days.[37] The same year, a 23-year-old gamer in Shanghai, China, died from exhaustion after playing World of Warcraft for 19 hours straight at an internet café.[38] There are also reports of gamers attacking and even shooting others when their systems were shut off or damaged.[39]

Of course, it's tempting to blame computer games for all sorts of societal maladies—from antisocial behavior to murder. In reality, they are a microcosm of society. Many individuals with mental health issues would likely step outside societal norms with or without technology in their lives. Yet, there's also evidence to suggest that behavioral problems can be worsened by the technology. The World Health Organization (WHO) updated its International

Classification of Diseases (ICD) codebook in early 2018 to include gaming disorders. It includes addiction as a problem when a person can't control the number of hours he or she is at a console, when it encroaches on other areas of life, and when a person can't stop—even to the point that it threatens a job or relationship.[40]

Finally, there are growing worries about people acting in increasingly antisocial ways. This includes an inability to distinguish between elements or events taking place in real and virtual worlds, and a more general problem of people treating each other badly inside virtual space and in the physical world. Just as people are more emboldened and act more aggressively behind the wheel of a vehicle, VR players may act more aggressively in a game or social space. Similarly, there are concerns that virtual worlds may lead to a decline in social skills—especially for those who spend a lot of time in VR spaces. While a person may not intend to be rude or mean, he or she may not recognize that certain behaviors cross a line.

Although there are no clear answers to all these questions, a pair of German researchers from Johannes Gutenberg University reported in 2016 that VR may create new ways to manipulate thinking.[41] This could introduce new challenges in areas as diverse as advertising, false news, and political propaganda. "VR poses risks that are novel, that go beyond the risks of traditional psychological experiments in isolated environments, and that go beyond the risks of existing media technology for the general public," the authors noted. Finally, they said, long-term immersion could trigger other unknown reactions, good and bad. One negative consequence could be depersonalization-derealization disorder, a psychological condition that revolves around a detachment from one's self.

The effects of this condition can be severe. "Extended time in VR could disrupt our normal experience of the real world as the real world, or attenuate the feeling of control that users have over their physical bodies and replace it with a stronger feeling of control over their avatar," Michael Madary and Thomas Metzinger wrote at LS:N Global, a news site devoted to consumer behavior.[42] "Since immersive technology is unregulated, it will be important for content providers to take the possibility of negative psychological consequences seriously."

Future Tense

Sorting through mountains of ethical and moral questions—and establishing a social and legal framework for dealing with extended reality—will not be easy. Many researchers have suggested an ethical code of conduct for researchers and companies.[43] This might include weighing the specific risks of VR exposure, integrating safeguards into applications, and introducing transparency into how data is collected and used, particularly when it's applied to scientific research. Rizzo says that, ultimately, developers must plan for the unexpected. "Because VR is continuing to evolve, not all of the questions regarding the consequences of VR exposure have been answered. As such, a thorough evaluation of all possible negative reactions should be conducted prior to initiation of protocol," he wrote.[44]

Many of these issues will remain in the spotlight for years to come—and take on different contexts for different uses and situations. There is a common theme, however. Depending on how people act, react, and interact, XR technologies could lead to radically different outcomes for both individuals and society. *Will they live second lives and have relationships in virtual worlds? Will people work in virtual worlds? And how will AR, VR, and MR fashion the thinking of young people?* Just as radio, television, the internet, and mobile phones shaped the values and sensibilities of generations and how they view and approach the world, extended reality will alter the course of human history. It's a future that's both enchanting and frightening.

7

EMBRACING AN AUGMENTED AND VIRTUAL FUTURE

Human and Machine Connect

The question isn't whether extended reality will become a standard part our world; it's *how*, *when,* and *where*. Another X-factor is what form these technologies will take and how society will adapt. Somewhere between utopian and dystopian outcomes lies the real world of adoption, benefits, costs, problems, and unintended consequences. It is clear that as VR, AR, and MR take shape, life will never be the same. XR will change the way we handle tasks, the way we buy things, the way we interact with others, and, ultimately, the way we think and behave.

It's clear that interest in virtual reality is high—and it continues to grow. According to a survey conducted by research firm Statistica, 77 percent of individuals between the ages of 20 and 29 are somewhat to very interested in

The question isn't whether virtual and augmented realities will become a standard part of our world; it's how, when, and where. Somewhere between utopian and dystopian outcomes lies the real world of adoption, benefits, costs, problems, and unintended consequences.

virtual reality. Among those between 30 and 39 years, the figure is 76 percent. Remarkably, 50 percent of those in the 40 to 49 age group are equally enthusiastic. Perhaps the most telling statistic of all is that only 4 percent of respondents between the ages of 14 and 19 indicated that they had no interest in virtual reality.[1] The remaining 96 percent are the people who will be shaping society over the next few decades.

What will augmented reality and virtual worlds look like in 10 or 20 years? How will we use these technologies and these virtual spaces in everyday life? How will they change the way we work? And how will they alter the way we communicate and interact with one another? There are no clear answers. In the end, perhaps only one thing is certain: the line between the physical world and the virtual world will increasingly blur and, in some cases, become completely indistinguishable. Humans and machines will become more interrelated and interconnected than ever before.

The Future Arrives

Extended reality is advancing at a furious pace. Every day, new concepts and applications emerge. As digital technologies converge further—and as software designers and developers learn how to create more compelling and useful apps and tools—AR, VR, and MR are moving from the fringes of society, something that mostly falls into the orbit of early adopters and tech geeks, to the center of business and life. Like mobility, the Internet of Things, artificial intelligence, and many other digital technologies, it will usher in massive social changes. Our homes, businesses, and public spaces will assume different shapes and forms than in a strictly physical world.

How will life change in the future? Virtual reality, augmented reality, and mixed reality will certainly not replace physical reality, but these technologies will introduce an overlay that sometimes displaces and often extends the "real" world. Designers, architects, scientists, lawyers, writers, salespeople, consultants, teachers, and many others will likely wear VR goggles to do some or all of their work—whether it's conducting research or creating a 3D computer model of a product or a building. People will also meet up in virtual spaces—and combine them with actual physical spaces—to create mixed-reality environments that assemble people, places, and things in new and different ways. In some cases, these spaces will introduce experiences that can only take place in an immersive digital world.

For example, an engineering firm might invite a developer to participate in a virtual tour of a proposed high-rise building. Meeting participants—scattered across different firms and all over the globe—would use an HMD to participate in a walk-through. Along the way, the facilitator might pull up technical specifications that display data overlays for physical objects and components. Participants would view the information and make possible modifications in real time. In the end, a design process that would have involved complicated travel schedules and coordinating meetings for dozens of people would be handled quickly, efficiently, and inexpensively. An added bonus: these gains would multiply over the course of a project.

The benefits of extended reality will extend to numerous fields—and include other digital tools. For instance, doctors, nurses, and technicians might rely on virtual environments to diagnose conditions and treat patients. The old-fashioned office visit could take place in a virtual space but include holograms that allow a physician to remotely view and diagnose a patient. Within this mixed-reality space, the doctor might use connected remote monitoring devices—blood pressure monitors, heartrate monitors, and more—to grab data about the patient or show the patient how a procedure or therapy works through 3D video or graphics.

Throughout the working world, it's possible to imagine entirely new ways for businesses to simplify processes and provide vastly better experiences for participants.

However, XR technologies will also change our lives outside the workplace. Within a decade, it's likely that online shopping sites will offer immersive virtual browsing and shopping. These virtual stores will likely resemble a physical store but also offer features that aren't possible in a physical space: instant access to information about products and even videos and holograms showing how they work. Selecting items will be a point-and-click proposition. Concurrently, physical stores will adopt elements of virtual storefronts. Shoppers wearing AR glasses will view product information and specs as they examine actual items.

XR technologies will also change the way we watch movies and sports, how we cook meals, experience the news, and use social media. They will provide a more intuitive way to install a light switch or assemble a piece of furniture. With our hands free and a recipe or instructional video projected onto AR glasses, digital and physical realms will blend into a hybrid reality that extends basic human functionality. In this space, there's no need to start and stop a video or tutorial on a phone or tablet. There's no need to pick up and set down a screwdriver or spatula along the way. Along with voice commands and gestures, it's possible to reduce friction points and the resulting frustration.

XR technologies will also change the way we watch movies and sports, how we cook meals, experience the news, and participate in social media. They will provide a more intuitive way to install a light switch or assemble a piece of furniture. With our hands free and a recipe or instructional video projected onto AR glasses, it's possible to blend digital and physical realities into a hybrid reality that extends basic human functionality.

Privacy: The Next Generation

The battle over privacy extends back to the origins of computing and digital technology. Of course, in recent years, the topic has attracted growing scrutiny. According to a 2018 Pew Research Center study, 93 percent of Americans believe that it's important to be in control of their data.[2] Another 2017 survey conducted by Pew found that 49 percent of Americans believe that their personal data was less secure compared to five years ago.[3] It's not difficult to understand why people are so anxious about cybersecurity. Breaches, breakdowns, and data theft have become daily events in today's world. What's more, they have very real consequences.

Extended reality will ratchet up the stakes. Already, web browsers, smartphones, GPS, and an assortment of clickstream and other data—including a person's purchase history with credit cards or through a loyalty program—provide a remarkably detailed picture of who someone is and what he or she does on a daily basis. XR, which can track a user's body movements, eye motion, and actual choices in a virtual environment, has the potential to deliver highly detailed psychological and psychographic profiles. It may also be possible to spot certain health problems though detailed monitoring. As virtual environments become more sophisticated, there's a risk that marketers and others could use private data in inappropriate, if not malicious, ways.

It's not an abstract concept. In 2016, Oculus released a privacy policy statement that revealed the company would collect data about users, including "information about your physical movements and dimensions when you use a virtual reality headset." The announcement triggered a public backlash.[4] In 2018, as a result of the European Union's General Data Protection Regulation (GDPR), which imposes stricter privacy standards for European citizens, Oculus updated its policy by introducing a privacy center where users can review data collected about them and make choices about what data is used.[5] Nevertheless, the XR privacy wars have only begun. There are concerns that VR could lead to increased identity theft as well as new types of government surveillance.

Augmented reality serves up a separate but equally vexing set of problems. For instance, when people began using Google Glass devices in 2013, the privacy backlash started almost immediately. The glasses could snap pictures, capture video, and broadcast scenes live to the world. Businesses, military instillations, and many others voiced concern about the use of devices to capture sensitive or secret information. Some went so far as to ban Google Glass. The list includes nightclubs, restaurants, and gyms that feared users would secretly capture and post images, including in private places like toilets and dressing rooms.

The Personal View

In the gaming world it's common to use avatars to represent a person or an object. The virtual world is likely to expand on this concept—while introducing entirely different ways for people to interact. This could include video and holographic avatars. It could include different avatars that manage different tasks or assist with different events. How all of this will change behavior and social frameworks remains to be seen. Thomas Metzinger, a philosopher at Johannes Gutenberg University in Mainz, Germany, believes that all of this points to a need to establish a code of conduct for virtual reality. "One's environment can influence one's psychological states," he wrote in a 2016 academic paper.[6] "Immersion in VR can have psychological effects that last after leaving the virtual environment."

Metzinger pointed out that usage issues revolve around four critical concerns: long-term immersion, neglect of the social and physical environment, risky content, and privacy. "VR is a technology, and technologies change the objective world," he noted. At the center of everything is the fact that objective changes are often subjectively perceived. This "may lead to correlated shifts in value judgments," he said. For example, virtual reality—particularly full-body experiences—will likely disrupt the way humans view the natural world and the relationship that people have with it.

Consider: a virtual image doesn't have to match its physical counterpart in order for a person to assume it's real and act upon the perceived reality. Research shows that when people look through goggles and view a live video feed of their own bodies (or a virtual body) located a short distance in front of their actual location they think of the image as their own body. Metzinger pointed out that when people viewed a virtual scene where someone scratched their back, they perceived that they were actually being stroked on their back; they couldn't differentiate between the two. In fact, the subjects reported feeling as though the virtual body was their own. Similarly, the manipulation of certain signals, including heartbeat, can influence how people think—and how they view the surrounding virtual environment.

Jeremy Bailenson at Stanford University's Virtual Human Interaction Lab believes that XR could enhance reality. "You can actually make VR better than face-to-face. You can make it hyper-personal or more social than you can have in the real world and there's a lot of ways to think about that ... with avatars you're always the way you want to look. You never have gestures that are inappropriate because you can have algorithmic filters that solve that. You can actually be doing very strange things, for example looking someone in the eye but looking two people in the eye at once and each of them feel like ... they're the only ones getting your eye contact."[7]

It's reasonable to assume that some people will opt out of the natural world whenever possible. These individuals may begin to think in patterns more like the virtual world than the physical world. Virtual reality, however, may also allow people to explore life as a different race, religion, or gender—or feel what it was like to live in a different era. "The fact that VR technology can induce illusions of embodiment is one of the main motivations [into understanding] the new risks generated by using VR by researchers and by the general public. Traditional paradigms in experimental psychology cannot induce these strong illusions," Metzinger noted. The question, of course, is whether these virtual spaces will enhance or undermine actual relationships.

The concept was at the center of the 2013 movie *Her*. It explored the idea of a man named Theodore Twombly (Joaquin Phoenix) establishing a romantically tinged relationship with a highly intelligent computer operating system embodied through a female voice named Samantha (Scarlett Johansson).[8] A depressed and soon-to-be divorced Twombly becomes increasingly infatuated by Samantha—and dependent on her to manage his daily life. Initially, Samantha supplies the validation and emotional support he finds severely lacking in human interactions, but then cracks develop. Imagine such a relationship in a virtual space. It's not difficult to envision that some people would prefer a simulation that delivers visual and tactile stimulation as a surrogate for real-world encounters and sex.

The World View

How VR will affect the individual isn't the only consideration. *How will the world interact differently? How will VR affect attitudes about other countries and cultures?* While there are no clear answers, it's also apparent that there's potential for deep behavioral manipulation, Metzinger points out. "Importantly, unlike other forms of media, VR can create a situation in which the user's entire environment is determined by the creators of the virtual world, including 'social hallucinations' induced by advanced avatar technology. Unlike physical environments, virtual environments can be modified quickly and easily with the goal of influencing behavior."

This has implications and ramifications that extend into religion, politics, and government. It also affects the news media—along with social media—which face a growing challenge to curb false news and the spread of inaccurate information. Metzinger says that virtual spaces change human thinking and the repercussions ripple into the physical world—influencing decisions large and small. In fact, academic research backs up his contention. For example, a 2014 study found that those who were deemed a "superhero" and given "superpowers" in the virtual world tended to act more altruistically, but those given traits of "villains" acted more malevolently.[9]

There are real-world implications for this concept. A separate study conducted by Spanish researchers found that when light-skinned individuals entered immersive virtual-reality environments and were given a dark-skinned body, their implicit racial bias decreased.[10] *Could such tools be used in schools to help young people—and adults for that matter—become more familiar with other races, religions, and cultures? Or could these same tools be used to fuel hatred and bigotry on a world scale?* In today's environment—where competing interests fight for the hearts and minds of the masses—the answer is likely to be both.

It's also conceivable that virtual gangs, militias, and military forces will emerge in virtual spaces. This may lead to entirely different ways to monitor, police, and protect participants. Finally, there are concerns that terrorists could use virtual worlds to inflict pain and suffering on others. This might take the form of horrific images, new and more serious types of malware threats, and the complete hijacking of virtual spaces. This raises additional questions: *Will there be a need for virtual police or vigilante forces that operate in these spaces? Will there be a need for punishments and prisons in virtual worlds? Will certain people be barred from certain areas of virtual worlds?* The ultimate question is: *How closely will the virtual world mirror the physical world and how will it differ?*

Out of the Body and into the Mind

The ability to alter the way people and society think is likely to filter into other aspects of life. Marketing and advertising have become far more sophisticated in recent decades. Political advertising has also advanced and taken on new dimensions. Researchers and data scientists have learned how to target people through the use of psychographic techniques that focus on personality, values, opinions, attitudes, and interests.

This includes the ability to micro-target advertising based on more granular demographics and characteristics. For example, an advertiser might create different ads with different phrases and language—including regional dialects—to appeal to different groups. Within the virtual world, this capability could extend to different races, religions, political orientations, and other factors. People's skin color, speech patterns, and behavior might vary in a program or in an advertisement. The application could even generate different avatars based on a user's desires—or biases.

In fact, virtual-reality advertising has already arrived. Adidas has created an ad to promote its Terrex outdoor gear by simulating rock climbing in an environment called Delicatessen VR.[11] Coca-Cola has produced a virtual-reality sleigh ride for the Oculus Rift,[12] in which a rider passes through a forested landscape, then takes off to fly over different villages. In the future, the feedback systems built into virtual reality could lead to deeper insights into how people are reacting to various stimulus in real time, Accenture's Viale says. This includes eye and head movements, where people choose to go inside the VR space and more. Advertisers could use this data to build "heat maps" of behavior—essentially, the common characteristics—but also adapt and adjust advertising in real time—and even personalize it on a one-to-one basis.

Psychographic data has profound ramifications that extend to the physical world. "You can imagine," Viale says, "that a person walks into a virtual shop and begins to look at things. What do they look at when they first walk in? How do they turn their head? What attracts their attention? What products or things do they tend to gravitate toward?" This data—collected from the sum of body movements—might determine how to better build virtual spaces as well as physical stores. Yet, this data could also be used by some to manipulate people for financial or political gain. "The data collected from virtual environments could allow people to understand the human mind like never before," Viale says.

A Day in the Life: 2030

Augmented reality, virtual reality, and mixed reality are advancing at a furious pace. Here's what a typical day might look like in 2030:

It's 7 a.m. on a Tuesday morning. Marc Smith wakes up and places his augmented-reality glasses over his eyes. He taps a button on the front of the frame and views a list of messages sent by people on his VIP list. He still uses a smartphone, but it has been relegated to a secondary role. The features and capabilities of the AR glasses now exceed that device. The glasses accept voice commands and display relevant and contextual data about places and things as Marc steps through his day.

Marc takes a shower, gets dressed, and heads into the kitchen for a quick breakfast. The AR glasses use infrared technology to scan his wheat toast and scrambled eggs and provide a calorie count based on the volume of the food. The system keeps track of his calories and nutrition throughout the day and he receives an alert if he's veering off his diet plan. While eating, he receives a text message from a client. He views it and then dictates a short response. If he needs to type out a longer message, he can switch on a holographic display that projects the image of a keyboard on the table. He can then tap out a note.

After breakfast, Marc uses the glasses to order a rideshare vehicle, and he continues to review messages and other documents. The glasses recognize that he is in a vehicle and thus provide contextually relevant information on request, including how long the commute to his office in downtown Seattle will take. When he finally catches up on incoming and outgoing messages, Marc watches the news, which is projected onto his glasses. He wears an earbud that delivers the audio from the video clips.

When Marc arrives at his office, he immediately prepares for a conference call with a client. As a senior project manager for a multinational engineering firm, it's his job to ensure that a major building project is going according to plan. He straps on a virtual-reality headset and uses a laser pointer in the virtual space to enter a group meeting. Marc connects to four others who are collaborating on the design of a new high-rise building in Barcelona, Spain. The other meeting participants are located in New York, Tokyo, Madrid, and Barcelona.

Once inside the virtual meeting place, Marc views avatars that resemble the other meeting participants. When a person speaks, his or her avatar illuminates while others dim. The room is filled with virtual display screens—think of these as a series of television screens—that allow Marc and the others to navigate through renderings, technical information, and more. He uses a laser pointer to select a screen and watch images, video, graphics, and data transform into an immersive 3D presentation.

Marc selects a model of the proposed structure. The system instantly drops all the participants into the virtual space. The group strolls through the entry and lobby, rides an elevator, and views how the various spaces will look when the project is completed. Marc shows a space that's designed for a restaurant. With the touch of a button, he changes the scene from an empty space to a bustling eatery. He then leads the group through office spaces and luxury apartments. At various points along the tour, he pulls up specs and details that project over the virtual space. Other meeting participants can also use their laser pointers to select objects and data. Twenty minutes later Marc concludes the tour. After a brief discussion all the participants sign off. Marc removes his VR headset and places the AR glasses over his eyes. A retina scan logs him into his desktop PC.

A few months later, when the building is actually under construction, Marc will visit the job site and, using AR glasses, tour the physical structure while viewing augmented overlays that display layers of information. This will include critical data about electrical systems, plumbing and heating, ventilation, and air conditioning (HVAC). For now, though, there's other work to do. So Marc steps into a virtual-reality holodeck at this office. He places a high-definition VR headset on his head and uses gestures and special gloves to view engineering and design options. He can quickly review different ideas and modifications clients have while gaining a realistic view of how the changes would appear in the physical world.

At 6 o'clock in the evening, Marc heads to the gym before going home. He hails a rideshare car and watches a TV show during the 25-minute drive. The glasses automatically adjust to optimize the display for the lighting in the car. At the gym, the glasses show how many steps he has taken and how many calories he has burned on the treadmill. Connected IoT sensors woven into his workout clothing also flash hydration alerts and other important information, including cues about when to push harder and when to ease up. When the sensors determine that he has reached his exercise goal the glasses flash a quick message congratulating him. He grabs a drink of water and heads for the shower.

A few minutes later, Marc stops at the grocery store to pick up a few items. His AR glasses allow him to view nutritional data about products. Since Marc wants to lose another 10 pounds, he's vigilant about the items he's buying and eating. When he holds a package or can in hand, the glasses scan the label and, if necessary, the barcode. Using image-recognition technology, the device transmits the data to a database in the cloud. A fraction of a second later, the AR glasses display information about the product. When he's finished shopping, Marc taps a button on the glasses. This activates a biometric eye scanner that authenticates and pays for the items. On the ride home, he video chats with his daughter, who is off at college.

After Marc arrives home and has dinner, he grabs his VR headset and a haptic body suit and fastens them. He browses a virtual reality version of the Web for a few minutes, and then pulls up a virtual-reality brochure for a resort in Tahiti. The HMD and body suit together create the feeling of actually being there, including the feeling of the sun, sand, and surf. Marc can check out the resort, step inside the rooms and overwater bungalows, view and smell the food, and even experience what it's like to snorkel among the fish and sea turtles in the lagoon. He's not ready to book the trip just yet, but if he were ready he could complete the purchase in the virtual space.

Finally, Marc selects a 3D adventure movie. The film offers a completely immersive experience that puts him at the center of the action but also lets him feel things that are taking place in the film. The body suit creates temperature and movement sensations—even while he is sitting on his sofa. It also allows him to experience tactile sensation when he touches an object. After watching the film, Marc swaps the HMD for AR glasses. He checks his messages again and prepares for bed. During the night, the glasses track his sleep patterns using a sensor located on a wristband he wears. During this time, the glasses sit on a recharging pad so they are ready for use in the morning.

A Virtual Future Unfolds

Reality always involves some level of perception. *Is a text document or an avatar residing inside a digital world really any different than a physical artifact? Does it have any less impact—and imprint—on the human mind than ink on a sheet of paper or an object that takes up space? Do people and things appearing in a virtual space have any less meaning than a person sitting across from us at a dining table?* At some point—and it's certainly not easy to recognize when—the physical and virtual worlds collide and morph into a broader single entity: *existence*.

Clearly, virtual reality, augmented reality, and mixed reality will shape and reshape our minds in the coming years. Extended reality will introduce new ways to work, play, and navigate the world. It will introduce entirely new types of applications and new forms of entertainment. The opportunities and possibilities are inspiring, but the pitfalls are also sometimes terrifying. XR has the potential to build bridges between people and things and make the world a better place. It has the potential to usher in enormous advances in learning, entertainment, and work. Yet, at the same time, these technologies are certain to create new and sometimes vexing challenges—along with new social problems.

Questions about XR strike at the very core of human existence. *Will these technologies change our lives and actually improve our world? Will they help people become more enlightened or will these technologies sink under the weight of flawed and often dysfunctional human behavior? Will XR introduce digital abstraction layers that lead us further away from nature? Will some people opt to live most of their lives in virtual reality at the expense of actual physical interaction? Could this lead to higher levels of loneliness, unhappiness, depression, and even suicide? And how will immersive virtual experiences affect the way we view other societies, religions, and people?*

In the end, one thing is certain: extended reality is here to stay. It's already entrenched in business and in life in ways we don't fully recognize. These connection points and exposure points will increase exponentially in the coming years. As virtual-reality pioneer Jaron Lanier writes in his book *Dawn of the New Everything, Encounters with Reality and Virtual Reality:* "You might have heard that VR failed for decades, but that was true only for the attempts to bring out a low-cost, blockbuster popular entertainment version. Just about every vehicle you've occupied in the last two decades, weather it rolls, floats, or flies, was prototyped in VR. VR for surgical training has become so widespread that concerns have been expressed that it's overused."

Ultimately, Lanier says, "The romantic ideal of virtual reality thrives as ever. VR the ideal, as opposed to the real, technology weds the nerdy thing with the hippie mystic thing; it's high-tech and like a dream or an elixir of unbounded experience all at the same time. ... Even though no one knows how expressive VR might eventually become, there is always that little core of thrill in the idea of VR."

Extended reality forces us to rethink basic assumptions about perception and the way we experience people, places, and things. If history is any indication—and there's no reason to think it isn't a good guide—the result will be a mélange of outcomes. Just as computers and other digital systems have transformed our world for better and for worse, augmented reality, virtual reality, and mixed reality will usher in a brave new world that fundamentally changes society. Strap on your AR glasses or HMD, put on your haptic gloves, and buckle up, because the voyage into augmented, virtual, and mixed realities is going to be one wild ride.

GLOSSARY

Application Programming Interface (API)
Software code that incorporates routines, protocols, and tools for connecting applications.

AR Glasses
Eyeglasses or goggles that serve as a digital interface between humans and computing devices. These devices augment the physical world with video, audio, and other forms of digital data.

Artificial Intelligence (AI)
Software algorithms that use a highly complex mathematical structure to boost computational decision making that approximates or exceeds human capabilities.

Aspect ratio
The screen proportion for viewing images in a virtual-reality or mixed-reality environment. If the aspect ratio is not properly set images will appear distorted.

Augmented Reality
The use of digital technology to enhance physical reality by displaying images, text, and other sensory data on a smartphone, glasses, windshield, or other media.

Avatar
An electronic or digital representation of an object or person from the physical world.

CAVE (Cave Automatic Virtual Environment)
A virtual-reality space that consists of a cube-shaped room with walls, floors, and ceiling serving as projections screens. The environment typically relies on a user wearing a head-mounted display and controllers to interact with virtual elements.

Central Processing Unit (CPU)
The circuitry inside a computing device that carries out the input/output (I/O) instructions to operate the device.

Data Glove
A device, often including haptics, that uses sensors to connect to a virtual-reality system. The gloves may incorporate hand movements and gestures that control the virtual space.

Dynamic Rest Frames
A technique that that adapts visual output dynamically to match the requirements of a virtual-reality experience. This helps reduce dizziness, nausea, and disorientation that can result in VR spaces.

Exoskeleton
A rigid covering for a body. In virtual reality, a suit that is worn over the body to provide tactile stimulation or sensation, to create the illusion of movement or power limbs.

Extended Reality (XR)
A term that refers to technologies that supplement or extend the physical world. This includes augmented reality (AR), virtual reality (VR), and mixed reality (MR).

Eye Tracking
Technology that uses sensors in a head-mounted display (HMD) to follow the motion of a user's eyes. This allows the system to adapt and adjust the display as necessary.

Field of View (FOV)
The field of vision in a virtual-reality, augmented-reality, or mixed-reality environment. FOV is measured by degrees. The greater the number the more lifelike the view.

Foveated Rendering
A technique that relies on eye tracking to deliver the highest quality images where the user is looking at any instant. The area outside the primary focus is rendered at much lower video quality.

Graphical Processing Unit (GPU)
Specialized circuitry that speeds graphical processing on computing devices.

Haptics
Technology that stimulates the senses of touch and motion, thereby creating a more realistic experience in virtual reality. This tactile feedback takes place through controllers, gloves, and full body suits, including exoskeletons.

Head-mounted Display (HMD)
A helmet or headset that includes a display component and other sensory input, such as audio.

Head Tracking
Technology that measures the user's head movements to adapt and adjust the visual display dynamically.

Heads-Up Display (HUD)
A projection system that uses augmented reality to generate data on a screen. This reduces or eliminates the need for the user to look away from the primary viewpoint.

Holodeck
A physical space that serves as the basic framework for a virtual space. A holodeck, which originated conceptually from *Star Trek*, typically consists of bare walls, floors, and ceilings that are transformed by a virtual-reality environment through an HMD. Using sensors, a person is able to walk through the space and view it as something entirely different.

Holography
A three-dimensional photographic image of an object that is projected into the physical or virtual worlds.

Immersive Virtual Reality
An artificial virtual environment that is entirely separated from the physical world.

Latency
A delay in visual or auditory output that results in a mismatch from other signals in the application or in the physical word.

Light-Emitting Diode (LED)
A type of display that uses a two-lead semiconductor as a light source. LED displays are commonly used in virtual reality.

Mixed Reality (MR)
A state of extended reality where elements of virtual reality and augmented reality appear together with physical objects.

Mixed-Reality Continuum
The range or spectrum of environments produced by combining virtual, augmented, and physical realities.

Omnidirectional Treadmill
A mechanical device that resembles a conventional unidirectional treadmill but allows 360-degree movement. In a VR space, this allows individuals to move around by foot, thus creating a more realistic environment.

Organic Light-Emitting Diode (OLED)
A light-emitting diode (LED) that uses an organic compound to emit light in response to an electric current. OLEDS are commonly used in televisions, computer monitors, mobile phones, and VR systems.

Polygons
The visual representation of data as displayed in augmented and virtual reality. The higher the polygon count the better the 3D representation, and the more realistic or immersive the virtual-reality space.

Proprioception
The human body's ability to orient itself using sensory input.

Refresh Rate
A technical specification that refers to how frequently images are updated or refreshed.

Simultaneous Localization and Mapping (SLAM)
A robotic mapping method that allows a computing system to map unknown or unpredictable elements while simultaneously keeping track of an agent's location within the space. The technology aids in orienting and rendering virtual environments.

Six Degrees of Freedom (6 DOF)
The flexibility of movement within a device or system. In VR, a device that possesses 6 DOF can move forward and backward, up and down, and left and right.

TrueDepth
A technology developed by Apple that uses infrared technology to measure and represent objects in 3D. The system is used for authentication as well as producing augmented reality elements.

Unity
A popular platform and API base for building virtual-reality environments

User Interface (UI)
The point of interaction between a computing device and a human. The UI provides the input and output required for interaction.

Virtual Reality (VR)
A computer-generated 3D environment that appears to be real or approximates the realities of the physical world.

NOTES

Introduction

1. "Virtual Reality Market Size Globally Reach US $26.89 Billion by 2022," Zion Market Research. https://globenewswire.com/news-release/2018/08/22/1555258/0/en/Virtual-Reality-Market-Size-Globally-Reach-US-26-89-Billion-by-2022-By-Major-Players-Technology-and-Applications.html
2. "Consumer AR Revenues to Reach $18.8 Billion by 2022," ARtillry, June 19, 2018. https://artillry.co/2018/06/19/consumer-ar-revenues-to-reach-18-7-billion-by-2022/
3. "Data Point of the Week: 50% More AR Companies in 2018," ARtillry, July 23, 2018. https://artillry.co/2018/07/23/data-point-of-the-week-50-more-ar-companies-in-2018/
4. "Consumer AR Revenues to Reach $18.8 Billion by 2022," ARtillry, June 19, 2018. https://artillry.co/2018/06/19/consumer-ar-revenues-to-reach-18-7-billion-by-2022/
5. "How Can Virtual Reality Plug in to Social Media," *Science Friday*, February 26, 2016. https://www.sciencefriday.com/segments/how-can-virtual-reality-plug-in-to-social-media/

Chapter 1

1. "View-Master," Wikipedia. https://en.wikipedia.org/wiki/View-Master
2. Living Wine Labels. https://www.livingwinelabels.com
3. "IKEA Launches IKEA Place, a New App that Allows People to Virtually Place Furniture in Their Home," IKEA, September 12, 2017. https://www.ikea.com/us/en/about_ikea/newsitem/091217_IKEA_Launches_IKEA_Place
4. Wikipedia. https://en.wikipedia.org/wiki/Telexistence
5. Merriam-Webster. https://www.merriam-webster.com/dictionary/augmented%20reality
6. Merriam-Webster. https://www.merriam-webster.com/dictionary/virtual%20reality
7. Shannon Selin, "Panoramas: 19th Century Virtual Reality," Shannon Selin: Imagining the Bounds of History. https://shannonselin.com/2016/11/panoramas-19th-century/
8. "Diorama," Wikipedia. https://en.wikipedia.org/wiki/Diorama

9. Stanley Grauman Weinbaum, "Pygmalion's Spectacles," *Project Gutenberg* https://www.gutenberg.org/files/22893/22893-h/22893-h.htm
10. "Link Trainer," *Encyclopedia Britannica*. https://www.britannica.com/technology/Link-Trainer
11. Stereoscopic Television Apparatus. Free Patents Online. http://www.freepatentsonline.com/2388170.html
12. Morton Heilig, "Inventor in the Field of Virtual Reality," mortonheilig.com. http://www.mortonheilig.com/InventorVR.html
13. Morton Heilig, Sensorama Patent. http://www.mortonheilig.com/SensoramaPatent.pdf`
14. "History of Virtual Reality: Where Did It All Begin?" Virtual Reality Guide. http://www.virtualrealityguide.com/history-of-virtual-reality
15. "Sketchpad: A Man-Machine Graphical Communication System." Internet Archive Wayback machine. https://web.archive.org/web/20130408133119/http://stinet.dtic.mil/cgi-bin/GetTRDoc?AD=AD404549&Location=U2&doc=GetTRDoc.pdf
16. Ivan E. Sutherland, "The Ultimate Display." http://www8.informatik.umu.se/~jwworth/The%20Ultimate%20Display.pdf
17. "The Sword of Damocles (virtual reality)," Wikipedia. https://en.wikipedia.org/wiki/The_Sword_of_Damocles_(virtual_reality)
18. Morton Heilig, Experience Theater Patent. http://www.mortonheilig.com/Experience_Theater_Patent.pdf
19. Heilig.
20. Caroline Cruz-Neira, Daniel J. Sandin, Thomas A. DeFanti, Robert Kenyon, and John C. Hart, "The CAVE: Audio Visual Experience Automatic Virtual Environment," *Communications of the ACM* 35, no. 6 (June 1992): 64–72. https://dl.acm.org/citation.cfm?doid=129888.129892
21. "Visbox." http://www.visbox.com/VisCube-models.pdf
22. "DSTS: First Immersive Virtual Training System Fielded." https://www.army.mil/article/84728/DSTS__First_immersive_virtual_training_system_fielded
23. DSTS.
24. Mark A. Livingston, Lawrence J. Rosenblum, Dennis G. Brown, Gregory S. Schmidt, Simon J. Julier, Yohan Baillot, J. Edward Swan II, Zhuming Ai, and Paul Maassel, "Military Applications of Augmented Reality.". https://www.nrl.navy.mil/itd/imda/sites/www.nrl.navy.mil.itd.imda/files/pdfs/2011_Springer_MilitaryAR.pdf
25. "Sega VR," YouTube, April 3, 2010. https://www.youtube.com/watch?v=yd98RGxad0U

26. "Waldern Virtuality," YouTube, October 31, 2016. https://www.youtube.com/watch?v=kw0-IKQJVeg&t=3m9s

27. "Apple Invents an Augmented Reality Windshield That Will Even Support FaceTime Calls between Different Vehicles," *Patently Apple*, August 4, 2018. http://www.patentlyapple.com/patently-apple/2018/08/apple-invents-an-augmented-reality-windshield-that-will-even-support-facetime-calls-between-different-vehicles.html

28. "Mars Immersion: NASA Concepts Bring Precision to New Virtual Reality Experience," NASA, December 16, 2015. https://www.nasa.gov/feature/nasa-concepts-bring-precision-mars-to-virtual-reality

29. London College of Communication. https://www.arts.ac.uk/subjects/animation-interactive-film-and-sound/postgraduate/ma-virtual-reality-lcc

30. Cole Wilson, "A Conversation with Jaron Lanier, VR Juggernaut," *Wired,* November 21, 2017. https://www.wired.com/story/jaron-lanier-vr-interview/

Chapter 2

1. Paul Milgram and Fumio Kishino, "A Taxonomy of Mixed Reality Visual Displays," *IEICE Transactions on Information Systems,* vol. E77-D, no. 12 (December 1994). http://etclab.mie.utoronto.ca/people/paul_dir/IEICE94/ieice.html

2. Augment. https://www.augment.com

3. Vanessa Ho, "Design Revolution: Microsoft HoloLens and Mixed Reality Are Changing How Architects See the Wworld," *Microsoft News*, June 2017. https://news.microsoft.com/transform/design-revolution-microsoft-hololens-mixed-reality-changing-architects-world/#sm.00000rxwyoih0gem6q2gakhajzadv

4. DAQRI. https://daqri.com/products/smart-glasses/

5. Alex Health, "Facebook Says the First Technology to Replace Smartphones Will Be Controlled with Our Brains," *Business Insider,* April 21, 2017. https://www.businessinsider.com/facebook-smart-glasses-will-be-controlled-with-our-brains-2017-4

6. "Celebrating 3 Million PS VR Systems Sold," PlayStation Blog. https://blog.us.playstation.com/2018/08/16/celebrating-3-million-ps-vr-systems-sold/

7. The CAVE Virtual Reality System. https://www.evl.uic.edu/pape/CAVE/

8. "Virtual Reality Becomes Reality for Engineering Students," March 5, 2018. http://www.lsu.edu/eng/news/2018/03/03-05-18-Virtual-Reality-Becomes-Reality.php

9. "Virtual Reality Becomes Reality for Engineering Students."

10. "A History of Haptics: Electric Eels to an Ultimate Display," HaptX Blog. https://haptx.com/history-of-haptics-electric-eels-to-ultimate-display/

11. Beimeng Fu, "Police in China Are Wearing Facial-Recognition Glasses," ABC News. February 8, 2018. http://abcnews.go.com/International/police-china-wearing-facial-recognition-glasses/story?id=52931801
12. "SF State Conducting Leading-Edge Research into Virtual Reality," *Fitness*, August 15, 2017. http://news.sfsu.edu/news-story/sf-state-conducting-leading-edge-research-virtual-reality-fitness
13. "SF State Conducting Leading-Edge Research into Virtual Reality."
14. Melena Ryzik, "Augmented Reality: David Bowie in Three Dimensions," *New York Times*, March 20, 2018. https://www.nytimes.com/interactive/2018/03/20/arts/design/bowie-costumes-ar-3d-ul.html
15. Graham Roberts, "Augmented Reality: How We'll Bring the News into Your Home." *New York Times* February 1, 2018. https://www.nytimes.com/interactive/2018/02/01/sports/olympics/nyt-ar-augmented-reality-ul.html
16. Medical Virtual Reality: Bravemind. http://medvr.ict.usc.edu/projects/bravemind/
17. Tanya Lewis, "Virtual-Reality Tech Helps Treat PTSD in Soldiers," *LiveScience*, August 8, 2014. https://www.livescience.com/47258-virtual-reality-ptsd-treatment.html
18. Jonathan Vanian, "8 Crazy Examples of What's Possible in Virtual Reality," *Fortune*, March 22, 2016. http://fortune.com/2016/03/22/sony-playstation-vr-virtual-reality/

Chapter 3

1. Colton M. Bigler, Pierre-Alexandre Blanche, and Kalluri Sarma, "Holographic Waveguide Heads-Up Display for Longitudinal Image Magnification and Pupil Expansion," *Applied Optics* 7, no. 9 (2018): 2007–2013. https://www.osapublishing.org/ao/ViewMedia.cfm?uri=ao-57-9-2007&seq=0&guid=8e7a4bd9-ac84-3c85-0db7-09016b8c3525&html=true
2. Augmented Reality Heads Up Display (HUD) for Yield to Pedestrian Safety Cues. Patent US9064420B2, 2015. https://patents.google.com/patent/US9064420B2/en
3. "Diffractive and Holographic Optics as Optical Combiners in Head Mounted Displays," UbiComp '13, September 8–12, 2013, Zurich, Switzerland. http://ubicomp.org/ubicomp2013/adjunct/adjunct/p1479.pdf
4. Meta. https://www.metavision.com
5. "Intel Vaunt," YouTube, February 13, 2018. https://www.producthunt.com/posts/intel-vaunt
6. MyScript. https://www.myscript.com

7. "Augmented Reality Being Embraced by Two-Thirds of Mobile Developers," Evans Data Corporation, October 4, 2017. https://evansdata.com/press/viewRelease.php?pressID=260
8. "The Real Deal with Virtual and Augmented Reality," Goldman Sachs, February 2016. http://www.goldmansachs.com/our-thinking/pages/virtual-and-augmented-reality.html
9. "The Real Deal with Virtual and Augmented Reality."
10. Anjul Paney, Joohwan Kim, Marco Salvi, Anton Kaplanyan, Chris Wyman, Nir Benty, Aaron Lefohn, and David Luebke, "Perceptually-Based Foveated Virtual Reality," Proceedings of SIGGRAPH 2016 Emerging Technologies, July 1, 2016. http://research.nvidia.com/publication/perceptually-based-foveated-virtual-reality
11. Doron Friedman, Christopher Guger, Robert Leeb and Mel Slater, "Navigating Virtual Reality by Thought: What Is It Like?" *Presence Teleoperators & Virtual Environments* 16, no. 1 (February 2007): 100–110, DOI: 10.1162/pres.16.1.100. https://www.researchgate.net/publication/220089732_Navigating_Virtual_Reality_by_Thought_What_Is_It_Like
12. Rachel Metz, "Controlling VR with Your Mind," *MIT Technology Review*, March 22, 2017. https://www.technologyreview.com/s/603896/controlling-vr-with-your-mind/
13. "How Brain-Computer Interfaces Work," *How Stuff Works*. https://computer.howstuffworks.com/brain-computer-interface.htm
14. "Touching the Virtual: How Microsoft Research Is Making Virtual Reality Tangible," Microsoft Research Blog, March 8, 2018. https://www.microsoft.com/en-us/research/blog/touching-virtual-microsoft-research-making-virtual-reality-tangible/
15. Activated Tendon Pairs in a Virtual Reality Device, United States Patent No. US 2018/0077976, March 22, 2018, p. 1. http://pdfaiw.uspto.gov/.aiw?PageNum=0&docid=20180077976&IDKey=C79DD0C718A0&HomeUrl=http://appft.uspto.gov/netacgi/nph-Parser%3FSect1%3DPTO1%2526Sect2%3DHITOFF%2526d%3DPG01%2526p%3D1%2526u%3D%25252Fnetahtml%25252FPTO%25252Fsrchnum.html%2526r%3D1%2526f%3DG%2526l%3D50%2526s1%3D%252522220180077976%252522.PGNR.%2526OS%3DDN/20180077976%2526RS%3DDN/20180077976
16. HoloSuit Kickstarter page. https://www.kickstarter.com/projects/holosuit/holosuit-full-body-motion-tracker-with-haptic-feed/description
17. "Google Plans Virtual-Reality Operating System Called Daydream," *Wall Street Journal*, May 18, 2016. https://www.wsj.com/articles/google-plans-virtual-reality-operating-system-called-daydream-1463598951

18. “This Treadmill Lets You Walk in Any Direction,” Engadget, May 20, 2014. https://www.engadget.com/2014/05/20/this-treadmill-lets-you-walk-in-any-direction/
19. Virtusphere. http://www.virtusphere.com
20. “First look at THE VOID,” YouTube, May 4, 2015. https://www.youtube.com/watch?time_continue=41&v=cML814JD09g
21. “Five Awesome VR Experiences to Try in Hong Kong,” *TimeOut*, August 17, 2018. https://www.timeout.com/hong-kong/things-to-do/five-awesome-vr-experiences-to-try-in-hong-kong
22. Sharif Razzaque, David Swapp, Mel Slater, Mary C. Whitton, and Anthony Steed, “Redirected Walking in Place,” EVGE Proceedings of the Workshop on Virtual Environments, 2002. https://www.cise.ufl.edu/research/lok/teaching/dcvef05/papers/rdw.pdf
23. “Walk on the Virtual Side,” How Stuff Works. https://electronics.howstuffworks.com/gadgets/other-gadgets/VR-gear2.htm
24. Paul Milgram, “A Taxonomy of Mixed Reality Visual Displays,” *IEICE Transactions on Information Systems*, Vol E77-D, no. 12 (December 1994). http://etclab.mie.utoronto.ca/people/paul_dir/IEICE94/ieice.html

Chapter 4

1. Wesley Fenlon, “The Challenge of Latency in Virtual Reality,” *Adam Savage’s Tested,* January 4, 2013. http://www.tested.com/tech/concepts/452656-challenge-latency-virtual-reality/
2. Sajid Surve, “What Is Proprioception?” BrainBlogger, June 9, 2009. http://brainblogger.com/2009/06/09/what-is-proprioception
3. J. Jerald, J.J. LaViola, R. Marks, Proceedings of SIGGRAPH ‘17 ACM SIGGRAPH 2017 Courses, Article no. 19, Los Angeles, California, July 30–August 3, 2017. https://dl.acm.org/citation.cfm?id=3084900&dl=ACM&coll=DL
4. Tom Vanderbilt, “These Tricks Make Virtual Reality Feel Real,” *Nautilus,* January 7, 2016. http://nautil.us/issue/32/space/these-tricks-make-virtual-reality-feel-real
5. J.L. Souman, L. Frissen, M.N. Sreenivasa, and M.O. Ernst, “Walking Straight into Circles,” *Current Biology* 19, no. 18 (September 29, 2009): 1538–1542, doi: 10.1016/j.cub.2009.07.053, Epub August 20, 2009. https://www.ncbi.nlm.nih.gov/pubmed/19699093
6. Jason Jerald, “VR Interactions,” SIGGRAPH 2017 Course Notes. http://delivery.acm.org/10.1145/3090000/3084900/a19-jerald.pdf?ip=73.37.62.165&id=3084900&acc=PPV&key=4D4702B0C3E38B35%2E4D4702B0C3E3

8B35%2EA2C613B6141B77EC%2E4D4702B0C3E38B35&__acm__=1534527790_26e3a9f2ae5be8e5d24f016e3bef7208
7. George ElKoura and Karan Singh, "Handrix: Animating the Human Hand," Eurographics/SIGGRAPH Symposium on Computer Animation (2003). https://www.ece.uvic.ca/~bctill/papers/mocap/ElKoura_Singh_2003.pdf
8. Benson G. Munyan, III, Sandra M. Neer, Deborah C. Beidel, and Florian Jentsch, "Olfactory Stimuli Increase Presence in Virtual Environments," *PLoS One*, 2016. https://www.ncbi.nlm.nih.gov/pmc/articles/PMC4910977/
9. Rachel Metz, "Here's What Happens When You Add Scent to Virtual Reality," *MIT Technology Review*, January 31, 2017. https://www.technologyreview.com/s/603528/heres-what-happens-when-you-add-scent-to-virtual-reality/
10. http://feelreal.com
11. Digital Lollipop. http://nimesha.info/lollipop.html#dtl
12. http://nimesha.info
13. Mark R. Mine, A. Yoganandan, and D. Coffey, "Principles, Interactions and Devices for Real-World Immersive Modeling," *Computers & Graphics* 29 (March 2015). https://www.researchgate.net/publication/273398982_Principles_interactions_and_devices_for_real-world_immersive_modeling
14. Neurable. http://www.neurable.com/about/science/neurable
15. Paul Debevec, Senior Researcher, "Experimenting with Light Fields: Google AR and VR," Google VR, March 14, 2018. https://www.blog.google/products/google-vr/experimenting-light-fields/
16. Jonathan Steuer, "Defining Virtual Reality: Dimensions Determining Telepresence," *Journal of Communication,* December 1992. https://onlinelibrary.wiley.com/doi/abs/10.1111/j.1460-2466.1992.tb00812.x

Chapter 5

1. Rachel Ann Sibley, "Trace3," *Vimeo*. https://vimeo.com/268744283
2. Chelsea Ekstrand, Ali Jamal, Ron Nguyen, Annalise Kudryk, Jennifer Mann, and Ivar Mendez, "Immersive and Interactive Virtual Reality to Improve Learning and Retention of Neuroanatomy in Medical Students: A Randomized Controlled Study." http://cmajopen.ca/content/6/1/E103.full.
3. Jacqueline Wilson, "Is Virtual Reality the Future of Learning? A New Study Suggests So," *Global News Canada,* March 22, 2018. https://globalnews.ca/news/4099122/virtual-reality-learning-study-cmaj-university-saskatchewan/
4. Erik Krokos, Catherine Plaisant, and Amitabh Varshey, "Virtual Memory Palaces: Immersion Aids Recall," *Journal of Virtual Reality,..* May 16, 2018. https://link.springer.com/article/10.1007%2Fs10055-018-0346-3

5. Washington Leadership Academy. http://www.washingtonleadership-academy.org/about/founding-story
6. "2015 Training Industry Report," *Training Magazine*. https://trainingmag.com/trgmag-article/2o15-training-industry-report
7. Sara Castellanos, "Rolls-Royce Enlists Virtual Reality to Help Assemble Jet Engine Parts," *Wall Street Journal*, September 21, 2017. https://blogs.wsj.com/cio/2017/09/21/rolls-royce-enlists-virtual-reality-to-help-assemble-jet-engine-parts/
8. "Virtual Reality Basketball Could Be Future of Sports Broadcasting," *San Francisco Chronicle*, April 5, 2017. https://www.sfchronicle.com/business/article/Virtual-reality-basketball-could-be-future-of-11053308.php
9. Crew Schiller, "Curry's Personal Trainer Working on 'Developing a Three-Minute Pregame Virtual Reality Drill.'" *NBC Sports*, June 14, 2018. https://www.nbcsports.com/bayarea/warriors/currys-personal-trainer-working-developing-three-minute-pregame-virtual-reality-drill
10. Al Sacco, "Google Glass Takes Flight at Boeing," *CIO Magazine*, July 13, 2016. https://www.cio.com/article/3095132/wearable-technology/google-glass-takes-flight-at-boeing.html
11. Deniz Ergürel, "How Virtual Reality Transforms Engineering," *Haptical*, October 14, 2016. https://haptic.al/virtual-reality-engineering-bd366c892583
12. DAQRI. https://daqri.com
13. "93 Incredible Pokemon Go Statistics and Fact," *DMR Business Statistics*, August 2018. https://expandedramblings.com/index.php/pokemon-go-statistics/
14. Paul Sawers, "Virtual Reality Movie Theaters Are Now a Thing," *VentureBeat*, March 5, 2016. https://venturebeat.com/2016/03/05/virtual-reality-movie-theaters-are-now-a-thing/
15. "Tokyo Cinemas to Show Virtual Reality Films for First Time," *Agencia EFE*, June 26, 2018. https://www.efe.com/efe/english/life/tokyo-cinemas-to-show-virtual-reality-films-for-first-time/50000263-3662112
16. Ben Pearson, "Paramount Has Created a Completely Virtual Movie Theater: Is This a Game Changer?" *Film*, November 17, 2017. https://www.slashfilm.com/virtual-reality-movie-theater/
17. "AltspaceVR Releases New Worlds and Custom Building Kits," AltspaceVR Blog. August 8, 2018. https://altvr.com/altspacevr-releases-worlds/
18. Statistica: The Statistics Portal. https://www.statista.com/statistics/499694/forecast-of-online-travel-sales-worldwide/
19. "Marriott Hotels Introduces the First Ever In-Room Virtual Reality Travel Experience," Marriott News Center. September 9, 2015. http://news.marriott

.com/2015/09/marriott-hotels-introduces-the-first-ever-in-room-virtual-reality-travel-experience/
20. Chris Morris, "10 Industries Rushing to Embrace Virtual Reality," CNBC, December 1, 2016. https://www.cnbc.com/2016/12/01/10-industries-rushing-to-embrace-virtual-reality-.html#slide=8
21. Smithsonian Journeys Venice. https://www.oculus.com/experiences/rift/1830344467037360/
22. Nikhloai Koolon, "Gatwick's Augmented Reality Passenger App Wins Awards," VR/focus, May 19, 2018. https://www.vrfocus.com/2018/05/gatwick-airportsaugmented-reality-passenger-app-wins-awards/
23. "CNNVR Launches on Oculus Rift," CNN, March 15, 2018. http://cnnpressroom.blogs.cnn.com/2018/03/15/cnnvr-launches-on-oculus-rift/
24. Dan Robitzski, "Virtual Reality and Journalistic Ethics: Where Are the Lines?" UnDark, September 27, 2017. https://undark.org/article/virtual-reality-and-journalistic-ethics-where-are-the-lines/
25. Paul Strauss, "Mini Augmented Reality Ads Hit Newsstands," Technabob, December 17, 2008. https://technabob.com/blog/2008/12/17/mini-augmented-reality-ads-hit-newstands/
26. Ann Javornik, "The Mainstreaming of Augmented Reality: A Brief History," *Harvard Business Review,* October 4, 2016. https://hbr.org/2016/10/the-mainstreaming-of-augmented-reality-a-brief-history
27. Sephora Virtual Artist. https://sephoravirtualartist.com/landing_5.0.php?country=US&lang=en&x=&skintone=¤tModel=
28. Matterport. https://matterport.com
29. Systems and Methods for a Virtual Reality Showroom with Autonomous Storage and Retrieval, United States Patent and Trademark Office, August 16, 2018. http://pdfaiw.uspto.gov/.aiw?docid=20180231973&SectionNum=1&IDKey=97E3E0213BD5&HomeUrl=http://appft.uspto.gov/netacgi/nph-Parser?Sect1=PTO1%2526Sect2=HITOFF%2526d=PG01%2526p=1%2526u=/netahtml/PTO/srchnum.html%2526r=1%2526f=G%2526l=50%2526s1=20180231973%2526OS=%2526RS=
30. "Lowe's Next-Generation VR Experience, Holoroom How To, Provides On-Demand DIY Clinics for Home Improvement Learning," March 7, 2017. https://newsroom.lowes.com/news-releases/holoroom-how-to/
31. Issy Lapowsky, "The Virtual Reality Sim That Helps Teach Cops When to Shoot," *Wired,* March 30, 2015. https://www.wired.com/2015/03/virtra/
32. Katie King, "Juries of the Future Could Be 'Transported' to Crime Scenes Using Virtual Reality Headsets," *Legal Cheek,* May 24, 2016. https://www.legalcheek.com/2016/05/juries-of-the-future-could-be-transported-to-crime-scenes-using-virtual-reality-headsets/

33. Megan Molteni, "Opioids Haven't Solved Chronic Pain. Maybe Virtual Reality Can," *Wired*. November 2, 2017. https://www.wired.com/story/opioids-havent-solved-chronic-pain-maybe-virtual-reality-can/
34. "An Augmented View," University of Maryland, October 12, 2016. https://www.youtube.com/watch?v=yDTjCYtr_4Y
35. AccuVein. https://www.accuvein.com
36. Surgical Theater. http://www.surgicaltheater.net
37. Precision Virtual Reality. https://www.gwhospital.com/conditions-services/surgery/precision-virtual-reality
38. The Virtual Reality Medical Center. http://vrphobia.com
39. "Entering Molecules," Medicine Maker, November 2016. https://themedicinemaker.com/issues/1016/entering-molecules/
40. Presley West, "How VR Can Help Solve Dementia," VR Scout, September 19, 2017. https://vrscout.com/news/vr-dementia/
41. "Virtual Reality May Help Students Experience Life with Dementia First Hand," *ScienceDaily*, July 23, 2018. https://www.sciencedaily.com/releases/2018/07/180723142811.htm
42. Tanya Lewis, "Virtual-Reality Tech Helps Treat PTSD in Soldiers," *LiveScience*, August 8, 2014. https://www.livescience.com/47258-virtual-reality-ptsd-treatment.html
43. Don Kramer, "New Simulators Get Stryker Drivers Up to Speed," US Army, April 27, 2007. https://www.army.mil/article/2881/new_simulators_get_stryker_drivers_up_to_speed
44. Brian Feeney, "Army's Augmented Reality Demo a Real Hit at the US Senate," US Army, November 16, 2016. https://www.army.mil/article/178491/armys_augmented_reality_demo_a_real_hit_at_the_us_senate
45. Claudui Romeo,"This Is the VR experience the British Army Is Using as a Recruitment Tool," *Business Insider*, May 28, 2017. http://www.businessinsider.com/british-army-virtual-reality-experience-recruitment-tool-challenger-tank-visualise-2017-5?r=UK&IR=T
46. Kristen French, "This Pastor Is Putting His Faith In a Virtual Reality Church," *Wired*, February 2, 2018. https://www.wired.com/story/virtual-reality-church/
47. Miranda Katz, "Augmented Reality Is Transforming Museums," *Wired*. April 23, 2018. https://www.wired.com/story/augmented-reality-art-museums/

Chapter 6

1. Robert Strohmeyer, "The 7 Worst Tech Predictions of All Time," *PC World*. https://www.pcworld.com/article/155984/worst_tech_predictions.html.

2. Strohmeyer, "The 7 Worst Tech Predictions of All Time."
3. *Experience on Demand. What Virtual Reality Is, How It Works, and What It Can Do* (W.W. Norton & Co. 2018), chapter 2.
4. "Health and Safety Warnings." *Oculus*. https://www.oculus.com/legal/health-and-safety-warnings/
5. Simon Worrall, "How Virtual Reality Affects Actual Reality," *National Geographic*, February 11, 2018. https://news.nationalgeographic.com/2018/02/virtual-reality-helping-nfl-quarterbacks--first-responders/
6. Jeremy Bailenson, *Experience on Demand: What Virtual Reality Is, How It Works, and What It Can Do*, Kindle Locations 1146–1150 (W. W. Norton & Co.), Kindle Edition.
7. Bailenson, *Experience on Demand*, Kindle Locations 1121–1122.
8. Arielle Michal Silverman, "The Perils of Playing Blind: Problems with Blindness Simulation and a Better Way to Teach about Blindness," *Journal of Blindness Innovation and Research* 5 (2015).
9. Bailenson, *Experience on Demand*, Kindle Locations 1161–1162.
10. Bailenson, *Experience on Demand*, Kindle Locations 1191–1195.
11. "Uncanny Valley," Wikipedia. https://en.wikipedia.org/wiki/Uncanny_valley
12. Craig Silverman and Jeremy Singer-Vine, "Most Americans Who See Fake News Believe It, New Survey Says," Buzzfeed News, December 6, 2016. https://www.buzzfeed.com/craigsilverman/fake-news-survey?utm_term=.laP5xgNB2#.fagOgkyxEs
13. "Violence in the Media," American Psychological Association. http://www.apa.org/action/resources/research-in-action/protect.aspx
14. C. A. Anderson, Nobuko Ihori, B. J. Bushman, H. R. Rothstein, A. Shibuya, E. L. Swing, A. Sakamoto, and M. Saleem, "Violent Video Game Effects on Aggression, Empathy, and Prosocial Behavior in Eastern and Western Countries: A Meta-Analytic Review," *Psychological Bulletin* 126, no. 2 (2010).
15. Patrick M. Markey and Christopher J. Ferguson, *Moral Combat: Why the War on Violent Video Games Is Wrong* (BenBella Books, 2017).
16. Joe Dawson, "Who Is That? The Study of Anonymity and Behavior," *Association for Psychological Science*, April 2018. https://www.psychologicalscience.org/observer/who-is-that-the-study-of-anonymity-and-behavior
17. L. Stinson and W. Ickes, "Empathic Accuracy in the Interactions of Male Friends versus Male Strangers," *Journal of Personality and Social Psychology* 62, no. 5 (1992): 787–797. http://dx.doi.org/10.1037/0022-3514.62.5.787
18. Stéphane Bouchard, Julie St-Jacques, Geneviève Robillard and Patrice Renaud, "Anxiety Increases the Feeling of Presence in Virtual Reality," *Presence:*

Teleoperators and Virtual Environments, 2008. https://www.mitpressjournals.org/doi/abs/10.1162/pres.17.4.376

19. David M. Hoffman, Ahna R. Girshick, Kurt Akeley, and Martin S. Banks, "Vergence–Accommodation Conflicts Hinder Visual Performance and Cause Visual Fatigue," US National Library of Medicine, National Institutes of Health, 2008. https://www.ncbi.nlm.nih.gov/pmc/articles/PMC2879326/

20. "Physical and Mental Effects of Virtual Reality," Tempe University Fox School of Business, 2016. http://community.mis.temple.edu/mis4596sec001s2016/2016/03/16/negative-physical-and-mental-affects-of-virtual-reality/

21. Sarah Griffiths, "Physical and Mental Effects of Virtual Reality: Could Oculus Rift KILL you? Extreme Immersion Could Become So Scarily Realistic It May Trigger Heart Attacks, Expert Warns," Daily Mail.com, August 26, 2014. https://www.dailymail.co.uk/sciencetech/article-2734541/Could-Oculus-Rift-KILL-Extreme-immersion-scarily-realistic-trigger-heart-attacks-expert-warns.html

22. Damon Brown, *Porn & Pong: How Grand Theft Auto Tomb Raider and Other Sexy Games Changed Our Culture* (Feral House, 2008), p. 21.

23. "Le Coucher de la Mariée," Wikipedia. https://en.wikipedia.org/wiki/Le_Coucher_de_la_Mariée

24. YouTube. https://www.youtube.com/watch?v=3R_sVeezhpY

25. Brown, *Porn & Pong,* p. 30.

26. Peter Rubin, "Coming Attractions: The Rise of VR Porn," *Wired,* April 17, 2018. https://www.wired.com/story/coming-attractions-the-rise-of-vr-porn/

27. Ashcroft v. Free Speech Coalition, *OLR Research Report,* May 3, 2002. https://www.cga.ct.gov/2002/rpt/2002-R-0491.htm

28. *The Nether: A Play*. https://www.amazon.com/dp/B00UPW4GOO?ref_=k4w_embed_details_rh&tag=bing08-20&linkCode=kpp

29. Olivia Solon, "Cybercriminals Launder Money Using In-Game Currencies," *Wired,* October 21, 2013. https://www.wired.co.uk/article/money-laundering-online

30. Mark A. Lemley and Eugene Volokh, "Virtual Reality, and Augmented Reality," *Law* (February 27, 2018); *University of Pennsylvania Law Review* 166 (2018); forthcoming; Stanford Public Law Working Paper no. 2933867; UCLA School of Law, Public Law Research Paper no. 17–13. https://ssrn.com/abstract=2933867 or http://dx.doi.org/10.2139/ssrn.2933867

31. Lemley and Volokh, p. 9.

32. Eugene Volokh, "Crime on the Virtual Street: Strobe Lighting, 'Virtual Groping,' and Startling," *Washington Post,* March 22, 2017. https://www.washingtonpost.com/news/volokh-conspiracy/wp/2017/03/22/crime-on-the-virtual-street-strobe-lighting-virtual-groping-and-startling/?utm_term=.510f841da86c

33. Lemley and Volokh, "Virtual Reality, and Augmented Reality," p. 10. https://ssrn.com/abstract=2933867 or http://dx.doi.org/10.2139/ssrn.2933867.

34. Jacqueline C. Campbell, "Physical Consequences of Intimate Partner Violence," 359 *The Lancet* 1331 (2002).

35. Tobias can Schneider, "The Post Virtual Reality Sadness," *Medium,* November 7, 2016. https://medium.com/desk-of-van-schneider/the-post-virtual-reality-sadness-fb4a1ccacae4

36. Frederick Aardema, Sophie Cote, and Kieron O'Connor, "Effects of Virtual Reality n Presence and Dissociative Experience," *CyberPsychology & Behavior* 9 (2006): 653–653. https://www.researchgate.net/publication/278206370_Effects_of_virtual_reality_on_presence_and_dissociative_experience

37. Katie Hunt, "Man Dies in Taiwan after 3-Day Online Gaming Binge," CNN, January 19, 2015. https://edition.cnn.com/2015/01/19/world/taiwan-gamer-death/index.html

38. Graham Reddick, "Chinese Gamer Dies after Playing World of Warcraft for 19 Hours," *Telegraph,* March 4, 2015. https://www.telegraph.co.uk/technology/11449055/Chinese-gamer-dies-after-playing-World-of-Warcraft-for-19-hours.html

39. "Side Effects: Death Caused by Video Game Addiction," *Deccan Chronicle,* January 10, 2016. https://www.deccanchronicle.com/150908/technology-latest/article/side-effects-deaths-due-caused-video-game-addiction

40. World Health Organization (WHO). ICD-11. 6C51. Gaming Disorder. https://icd.who.int/dev11/l-m/en#/http%3a%2f%2fid.who.int%2ficd%2fentity%2f1448597234

41. Michael Madary and Thomas K. Metzinger, "Real Virtuality: A Code of Ethical Conduct. Recommendations for Good Scientific Practice and the Consumers of VR-Technology," *Frontiers in Robotics and AI,* February 19, 2016. https://www.frontiersin.org/articles/10.3389/frobt.2016.00003/full

42. Michael Madary and Thomas Metzinger, "Virtual Reality Ethics," *LS:N Global*, April 4, 2016. https://www.lsnglobal.com/opinion/article/19360/madary-and-metzinger-virtual-reality-ethics

43. Michael Madary and Thomas K. Metzinger, "Real Virtuality: A Code of Ethical Conduct. Recommendations for Good Scientific Practice and the Consumers of VR-Technology." *Frontiers in Robotics and AI,* February 19, 2016. https://www.frontiersin.org/articles/10.3389/frobt.2016.00003/full
44. Albert "Skip" Rizzo, Maria T. Schultheis and Barbara O. Rothbaum, "Ethical Issues for the Use of Virtual Reality in the Psychological Sciences," *Ethical Issues in Clinical Neuropsychology* (Swets & Zeitlinger, 2003), 243–280. http://www.virtuallybetter.com/af/documents/VR_Ethics_Chapter.pdf

Chapter 7

1. "Are You Interested in Virtual Reality?" *Statistica: The Statistic Portal,* 2015. https://www.statista.com/statistics/456812/virtual-reality-interest-in-the-united-states-by-age-group/
2. Abigail Geiger, "How Americans Have Viewed Government Surveillance and Privacy since Snowden Leaks." *Pew Research Center,* June 4, 2018. http://www.pewresearch.org/fact-tank/2018/06/04/how-americans-have-viewed-government-surveillance-and-privacy-since-snowden-leaks/
3. Pew Research Center, "Americans and Cybersecurity," January 26, 2017. http://www.pewinternet.org/2017/01/26/1-americans-experiences-with-data-security/#roughly-half-of-americans-think-their-personal-data-are-less-secure-compared-with-five-years-ago
4. Will Mason, "Oculus 'Always On' Services and Privacy Policy May Be a Cause for Concern," Upload, April 1, 2016. https://uploadvr.com/facebook-oculus-privacy/
5. Adi Robertson, "Oculus Is Adding a 'Privacy Center' to Meet EU Data Collection Rules," The Verge, April 19, 2018. https://www.theverge.com/2018/4/19/17253706/oculus-gdpr-privacy-center-terms-of-service-update
6. Michael Madary and Thomas K. Metzinger. "Real Virtuality: A Code of Ethical Conduct. Recommendations for Good Scientific Practice and the Consumers of VR-Technology," Johannes Gutenberg, Universität Mainz, Mainz, Germany, February 19, 2016. https://www.frontiersin.org/articles/10.3389/frobt.2016.00003/full
7. "How Can Virtual Reality Plug in to Social Media? Interview with Jeremy Bailenson," Science Friday, February 2, 2016. https://www.sciencefriday.com/segments/how-can-virtual-reality-plug-in-to-social-media/
8. *Her*. Warner Bros. https://www.warnerbros.com/her
9. Gunwood Yoon and Patrick T. Vargas, "Know Thy Avatar: The Unintended Effect of Virtual-Self Representation on Behavior," Association for

Psychological Science, February 5, 2014. http://journals.sagepub.com/doi/abs/10.1177/0956797613519271?journalCode=pssa

10. Tabitha C. Peck, Sofia Seinfeld, Salvatore M. Aglioti, and Mel Slater, "Putting Yourself in the Skin of a Black Avatar Reduces Implicit Racial Bias," *Consciousness and Cognition* 22, no. 3 (September 2013): 779–787. https://www.sciencedirect.com/science/article/pii/S1053810013000597

11. Somewhere Else x Adidas: Delicatessen VR (Trailer). https://www.youtube.com/watch?time_continue=52&v=-1yhQF-rwi4

12. "Coca Cola Virtual Reality for Christmas," YouTube. https://www.youtube.com/watch?v=bTbfPALVQgs

FURTHER READING

Alter, Adam. *Irresistible: The Rise of Addictive Technology and the Business of Keeping Us Hooked*. Penguin Books, 2018.

Bailenson, Jeremy. *Experience on Demand: What Virtual Reality Is, How It Works, and What It Can Do*. Jeremy. W.W. Norton & Company, 2018.

Bucher, John. *Storytelling for Virtual Reality*. Routledge, 2017.

Carr, Nicholas. *The Shallows: What the Internet Is Doing to Our Brains*. W. W. Norton & Company, 2011.

Ceruzzi, Paul E. *Computing: A Concise History*. MIT Press, 2012.

Frankish, Keith. *The Cambridge Handbook of Artificial Intelligence*. Cambridge University Press, 2014.

Ewalt, David. *Defying Reality: The Inside Story of the Virtual Reality Revolution*. Blue Rider Press. 2018

Fink, Charlie. *Charlie Fink's Metaverse—An AR Enabled Guide to AR & VR*. Cool Blue Media, 2018.

Greengard, Samuel. *The Internet of Things*. MIT Press, 2015.

Grimshaw, Mark. *The Oxford Handbook of Virtuality*. Oxford University Press, 2013.

Jones, Lynette. *Haptics*. MIT Press, 2018.

Kline, Earnest. *Ready Player One*, Earnest, Broadway Books, 2012.

Kromme, Christian. *Humanification: Go Digital, Stay Human*. The Choir Press, 2017.

Kurzweil, Ray. *How to Create a Mind: The Secret of Human Thought Revealed*. Penguin Books, 2012.

Lanier, Jaron. *Dawn of the New Everything: Encounters with Reality and Virtual Reality*. Henry Holt and Co., 2017.

Lanier, Jaron. *Who Owns the Future?* Simon & Schuster, 2013.

Markey, Patrick M. and Ferguson, Christopher J. *Moral Combat: Why the War on Violent Video Games Is Wrong*. BenBella Books. 2017.

Metzinger, Thomas. *The Ego Tunnel: The Science of the Mind and the Myth of the Self*. Basic Books, 2010.

Papagiannis, Helen. *Augmented Human: How Technology Is Shaping the New Reality*. O'Reilly Media, 2017.

Rheingold, Howard. *Virtual Reality: The Revolutionary Technology of Computer-Generated Artificial Worlds-And How It Promises to Transform Society*. Simon & Schuster, 1991.

Rubin, Peter. *Future Presence: How Virtual Reality Is Changing Human Connection, Intimacy, and the Limits of Ordinary Life*. HarperOne, 2018.

Turkle, Sherry. *Alone Together: Why we Expect More from Technology and Less from Each Other*. Basic Books, 2017.

Weinbaum, Stanley G. *Pygmalion's Spectacles*. Project Gutenberg, 2007

Wiener, Norbert. *Cybernetics, Second Edition: or the Control and Communication in the Animal and the Machine*. Martino Fine Books, 2013.

Williamson, Roland. *A Virtual Agent in A Virtual World: A Brief Overview of Thomas Metzinger's Account of Consciousness*. Amazon Digital Services LLC, 2017.

INDEX

The MIT Press Essential Knowledge Series

Auctions, Timothy P. Hubbard and Harry J. Paarsch
The Book, Amaranth Borsuk
Carbon Capture, Howard J. Herzog
Cloud Computing, Nayan B. Ruparelia
Computational Thinking, Peter J. Denning and Matti Tedre
Computing: A Concise History, Paul E. Ceruzzi
The Conscious Mind, Zoltan E. Torey
Crowdsourcing, Daren C. Brabham
Data Science, John D. Kelleher and Brendan Tierney
Deep Learning, John D. Kelleher
Extremism, J. M. Berger
Food, Fabio Parasecoli
Free Will, Mark Balaguer
The Future, Nick Montfort
GPS, Paul E. Ceruzzi
Haptics, Lynette A. Jones
Information and Society, Michael Buckland
Information and the Modern Corporation, James W. Cortada
Intellectual Property Strategy, John Palfrey
The Internet of Things, Samuel Greengard
Machine Learning: The New AI, Ethem Alpaydin
Machine Translation, Thierry Poibeau
Memes in Digital Culture, Limor Shifman
Metadata, Jeffrey Pomerantz
The Mind–Body Problem, Jonathan Westphal
MOOCs, Jonathan Haber
Neuroplasticity, Moheb Costandi
Nihilism, Nolen Gertz
Open Access, Peter Suber
Paradox, Margaret Cuonzo
Photo Authentication, Hany Farid
Post-Truth, Lee McIntyre
Robots, John Jordan
School Choice, David R. Garcia
Self-Tracking, Gina Neff and Dawn Nafus
Sexual Consent, Milena Popova